PHOTOGRAMMETRY APPLIED TO ELECTRON MICROGRAPHY

By the Same Author
Fundamentals of Computational Photogrammetry

About the Author

Sanjib K. Ghosh worked as a Surveyor in the Survey of India for 11 years, and as a faculty member at the Ohio State University for 20 years and Laval University for 12 years. Also, he served the United Nations as a Consultant on World Cartography. His professional experience in research and development involved, among others: Air Photo Mapping of India - East Pakistan Boundary; Ice flow movement studies in Antarctica; Calibration of the Lunar Orbiter IV (for Moon mapping by NASA); Calibration of the world's largest radio telescope (at Arecibo, Puerto Rico); Mapping blood cells (Electron-micrographic); Calibration of Electron Microscopes; 3-D Mapping of human brain (Stereo-radiographic); and Self-calibration used in unconventional photogrammetric systems.

He also served as Visiting Professor at various institutes and universities in Brazil, Canada, India, Japan, Korea, Malaysia, Morocco, Nigeria and Thailand. He has authored around 200 scientific publications, four books as sole author and four books as a contributing author. Inventor of a New Method of Determining Latitude and Azimuth, Dr. Ghosh holds a patent on the use of 'Floating Line' in Stereo-radiographic surgery of epileptic human brains.

PHOTOGRAMMETRY APPLIED TO ELECTRON MICROGRAPHY

SANJIB K. GHOSH
Professor (Emeritus)
The Ohio State and Laval Universities

CONCEPT PUBLISHING COMPANY, NEW DELHI-110059

Cataloging in Publication Data—DK
Courtesy: D.K. Agencies (P) Ltd. <docinfo@dkagencies.com>

Ghosh, Sanjib K. (Sanjib Kumar), 1925-
Photogrammetry applied to electron micrography / Sanjib K. Ghosh.
p. cm.
Includes bibliographical references (p.)
Includes index.
ISBN 8180692744

1. Electron microscopy. 2. Photogrammetry.
I. Title.

DDC 502.825 22

ISBN : 81-8069-274-4 (H.B.)

First Published 2006

Published and Printed by
Ashok Kumar Mittal
Concept Publishing Company
A/15-16, Commercial Block, Mohan Garden
New Delhi-110059 (INDIA)

Phone : 25351460, 25351794
Fax : 091-11-25357103
Email : publishing@conceptpub.com

In Dedication

To

Swami Ranganathananda

[of the Ramakrishna Order]

I try my best to live up to his advice:

Give your best…
Whatever you do
Wherever you are,
Whenever you can.

PREFACE

Photogrammetry [the term credited to Kersten (1855 in German literature) and Meydenbauer (1867 in international literature)] has come a long way in 2005. The period between World Wars I and II (1916-1945) witnessed extensive use of aerial photographs for mapping and for military intelligence. Since its inception, the principal application of aerial photogrammetry has been in the production of topographic maps, through which it has been developed into a highly efficient measurement technique.

Its natural growth has been in applications outside the realm of topographic mapping, which, in a way, prompted the formation of Commission V [one of seven technical divisions devoted to non-topographic applications] of the International Society for Photogrammetry (ISP, currently ISPRS) at the Hague Congress in 1948. In the space age (since around 1955), the application developments have been in the Macro-range (with stellar cameras, satellite borne imageries, ballistic cameras, etc.) as well as in the Close-range (with photo-theodolites, stereometric cameras, radiography, etc.). Most recent developments have been in the Micro-range (with international encouragements through ISPRS Commission V).

In today's technology, sensor systems used in photogrammetry and remote sensing include, beyond the traditional camera (with film or glass) systems, in addition to X-rays, scanners with photomultipliers, solid-state detectors, radars, radiometers, lidars, scanning and transmission electron microscopes and others. The mathematical models in expressing the image formations comprise primarily the *central projection* (bundles), the *parallel projection* (beams) and *scanner-generated geometries* (lines or spots).

Electron Microscopes (EMs) have had a great impact on man's understanding of the structure and function of numerous biological and physical entities. Knowledges gained through the use of EMs have been derived mostly from qualitative (interpretative) two-dimensional analyses with relatively few quantitative (mensural) two- or three-dimensional studies. In extending such knowledges, attempts have been made during the last four decades to adopt and develop the quantitative techniques to their utmost. The use of modern image processing methods and available High-voltage Electron Microscopes (HVEMs) have greatly facilitated such analyses and have been found to be very powerful tools. The HVEMs are having increasing impacts on various ultrastructures. The information they provide is essential and unobtainable from other methods.

The author's background in photogrammetry (spatial, aerial, terrestrial and close-range) and his interest in expanding into micrographic areas through research have been the motivating forces in preparing materials for this book. In this regard, contributions made by his erstwhile student-advisee-associates [primarily Dr. Muzaffer Adiguzel of Turkey, Dr. Shawki El Ghazali of Egypt, Dr. David Maune of the USA and Dr. Hebbur Nagaraja of India] have been considerable. Their contributions are gratefully acknowledged.

The purpose of this book is to provide a basis for the use of two- and three-dimensional quantitative studies in EMs by applying the technology of photogrammetry. The approach is somewhat mathematical. However, its intent is to provide the necessary background for workers to whom something additional has always been felt desirable. The use of this book presupposes some knowledge of elementary college-level mathematics: linear, matrix and vector algebra; trigonometry; and projective geometry. For a greater appreciation, the reader should also understand the principles of least squares, propagation of variances, and their current symbology as used in the fields of engineering and physical sciences.

The book is divided into six chapters. A brief introduction of the rudimentaries of photogrammetric mensuration and essentials of the EM systems are presented in chapter 1. The second chapter discusses the geometric aspects in some details. Chapter 3 contains a brief discussion on all types of measuring instruments used in EM applications. Accuracy and reliability considerations at isolated points and in continuous plotting are next discussed in chapter 4. Application of the principles of photogrammetric mensuration to single specimens and ideas of combining data from both, SEM and TEM are discussed in chapter 5. Finally, in view of the currently available modern technologies of computer aided digital data processing and displaying, the digital terrain modeling ideas are presented with examples from real samples in chapter 6. The interested reader would benefit from further ideas, if felt needed, from the related literature cited in the bibliography in the end.

Numerous post-graduate students at The Ohio State University and at Laval University (Quebec) were exposed to the ideas presented in this book. Their reactions led to some modifications. I have not maintained an accurate record of who said what, their contributions must remain, therefore, anonymous—but they deserve the reader's gratitude. Finally, I thank my departed wife, Tapati, for her help, which was of a different nature but at least equally valuable. Her terminal illness caused some delay in finalizing the manuscript and in publishing this book. The superb job of publishing by the team at Concept Pulishing Company is appreciated.

Columbus, Ohio, USA
November 2005

Sanjib K. Ghosli

CONTENTS

LIST OF FIGURES AND TABLES

LIST OF TABLES

LIST OF ABBREVIATIONS AND ACRONYMS

2-D	Two-dimensional
3-D	Three-dimensional
AP/C	Analytical Plotter, Commercial
ASPRS	American Society for Photogrammetry and Remote Sensing
CRT	Cathode Ray Tube
DEM	Digital Elevation Model (also DTM)
DTM	Digital Terrain Model (also DEM)
DPW	Digital Photogrammetric Workstation
EM	Electron Microscope (or Micrography)
EMPD	Electron Micrograph Plotting Device
GIS	Geographic Information System
ISPRS	International Society for Photogrammetry and Remote Sensing
OLM	Optical Light Microscope
PC	Personal Computer
Pixel	Picture element
QA	Qasco
SEM	Scanning Electron Microscope
SIRA	Scientific and Industrial Research Association
SW	Softcopy Workstation
TEM	Transmission Electron Microscope
TV	Television

1

INTRODUCTION

Photogrammetry (Gk *phot* = light; Gk *gramma* = writing and Gk *metron* = to measure), strictly speaking, means the measurement of images on photographs. More generally, however, photogrammetry refers to the science and technology of obtaining reliable information about physical objects by recording, measuring and interpreting images. Broadly, in view of its applications, it can be considered in three categories, viz., Remote-range (Aerial, Terrestrial and Space), Close-range and Micro-range. This book is intended to deal with its application in the Micro-range, where, as an efficient tool of mensuration, photogrammetry has recently developed into a highly reliable and precise technique.

1.1 PHOTOGRAMMETRIC MENSURATION

There are two distinct aspects of photogrammetry: (1) *Quantitative* (or metric), which involves dimensional measurements to obtain *direct* information related to size and shape of objects; or *derived* information such as change in volume; statistical parameters like area distribution and standard deviation; or other associated parameters such as stress and force; and (2) *Qualitative* (or interpretative), which deals with the recognition and interpretation of objects. Visual, graphic or digital representation of such information is almost always implied. Our discussions will be limited to the 'quantitative' and 'direct' part, the rest being considered beyond the scope of this book.

Except by chance, a value obtained from a measurement is never exactly true. Although many authors (engineers or scientists) use the two words "measurement" and "observation" interchangeably, there exists a subtle difference between them. [***Note***: *Measurement* (Eisenhart, 1969) is the assignment of numbers to objects to represent the relations existing among them with respect to particular properties. In a quantitative sense, one does not measure an object but rather its dimensional properties like length, area and volume. *Observation* refers to both, the operational process and its outcome, especially with regard to the associated "adjustment" procedure]. In the context of both terms, the following considerations are necessary:

(a) Measurements are made with either simple or complex instruments or with the aid of physical phenomena (like periods of time or breaking of cells). Associated with these are the influences of human observers (operators) with their individual characteristics.

(b) Measurements should be referred to the corresponding *standards* with which direct or indirect comparisons are to be made. Associated with these are the corresponding dimensions and their units.
(c) The measurement involves a physical operation. Often, however, several other more operations are implied such as preparation (setting up, acceptation of known data, system calibration, etc.), pointing and comparing.
(d) The numerical data obtained from the observations represent the measurements. The data indicate the conditions and circumstances under which the measurements are made.
(e) Often the measurements are means to other ends (even within an instrument, like angles and distances are measured to obtain locations, areas, etc.) giving *derived* data—directly or indirectly.
(f) Such derived data are based on *theoretical concepts* (expressed by "mathematical models" based on physical laws). The possibility remains, however, that such concepts are not fully or correctly established.

Note: *Mathematical Models*: A physical situation or a set of events can be expressed often through the symbolic (abstract) language of mathematics. Symbolic expressions, known as *mathematical models,* provide the theoretical backgrounds. But they have no scientific value unless and until they have been adequately validated. Scientific validation is an open-ended processes. The *model* becomes established when it has been successfully tested for repeated applications. Otherwise, depending on the degree of failure, it stands to be changed, modified or simply rejected.

For using a mathematical model in practice, the expression should describe the situation as exactly as possible and it should be simple enough to provide solutions with no undue difficulties. These two requirements are not always compatible. Thus, one may have to compromise with the model to obtain acceptable results, with full knowledge that the results may not be exact (because the model thereby becomes approximate). In this regard, it is often convenient in practice to apply several equations sequentially, because such an approach is simple, is easy for detecting operational blunders and is applicable by the user with judgment (depending on the degree of exactness necessary for that specific application). However, a sequential approach also implies primary, secondary, tertiary, etc. effects of importance.

Mathematical models are composed of two parts: (1) The *functional model*, which presents the deterministic properties of the physical situation being considered. For precise quantitative work, usually a static (time-independent) situation is assumed. Dynamic mensuration and the related problems are beyond the scope of this book. (2) The *stochastic model*, which describes and defines the non-deterministic or probabilistic (stochastic) properties of involved variables that represent the measurements. These models are subject to statistical evaluations, which offer ways of estimating the quality of the work.

This book will cover basically the functional model. We would have, however, adequate consideration of the associated stochastic state. For ideas on stochastic models and corresponding principles of "adjustment computations", the reader may benefit from books on that subject (e.g., Mikhail, 1976 or Wolf and Ghilani, 1997). Established ideas on *Accuracy* and *Reliability* considerations will be given in a later chapter.

1.2 ELECTRON MICROSCOPES

Electron microscopes (EMs) have opened new opportunities to various applications in science and engineering. They are almost indispensable investigation tools in fields like metallurgy, microbiology, microelectronics, criminology, etc. Single, two-dimensional micrographs are often satisfactory for many uses. However, there is a growing demand for three-dimensional information. Many of these seem to be nothing but microscopic versions of macroscopic cases usually known to photogrammetrists. However, the projection geometry and inherent deformations are different in the EMs imaging systems. There are also certain limitations in the photogrammetric parameters and in providing "control" for EM imagery, where conventional approaches are usually not satisfactory enough.

EMs are basically of two types: Scanning Electron Microscope (SEM) and Transmission Electron Microscope (TEM). Representative examples from two well known manufacturers of EMs are presented in Figs. 1 and 2. They both, however, differ from the Optical Light Microscope (OLM) and from each other in several ways (see Table 1). In both SEM and TEM an electron beam is used for illumination and this beam impinges upon the specimen. While a SEM micrograph depicts the surface details of the object (specimen), a TEM micrograph depicts the inside of a thin specimen, somewhat similar to a X-ray radiograph. While surface mapping potential with SEM is great, TEM, with its larger magnification capability, gives a much better resolving power.

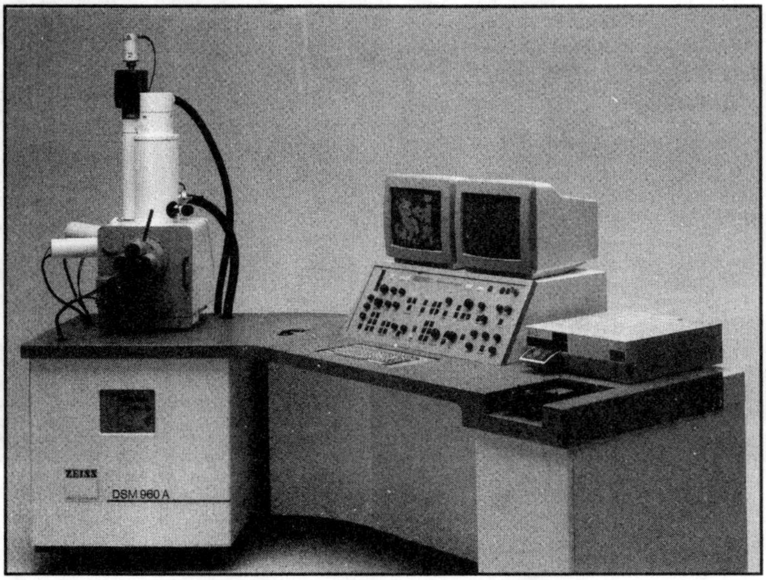

Fig. 1: Zeiss DSM 960 A, SEM (general view)

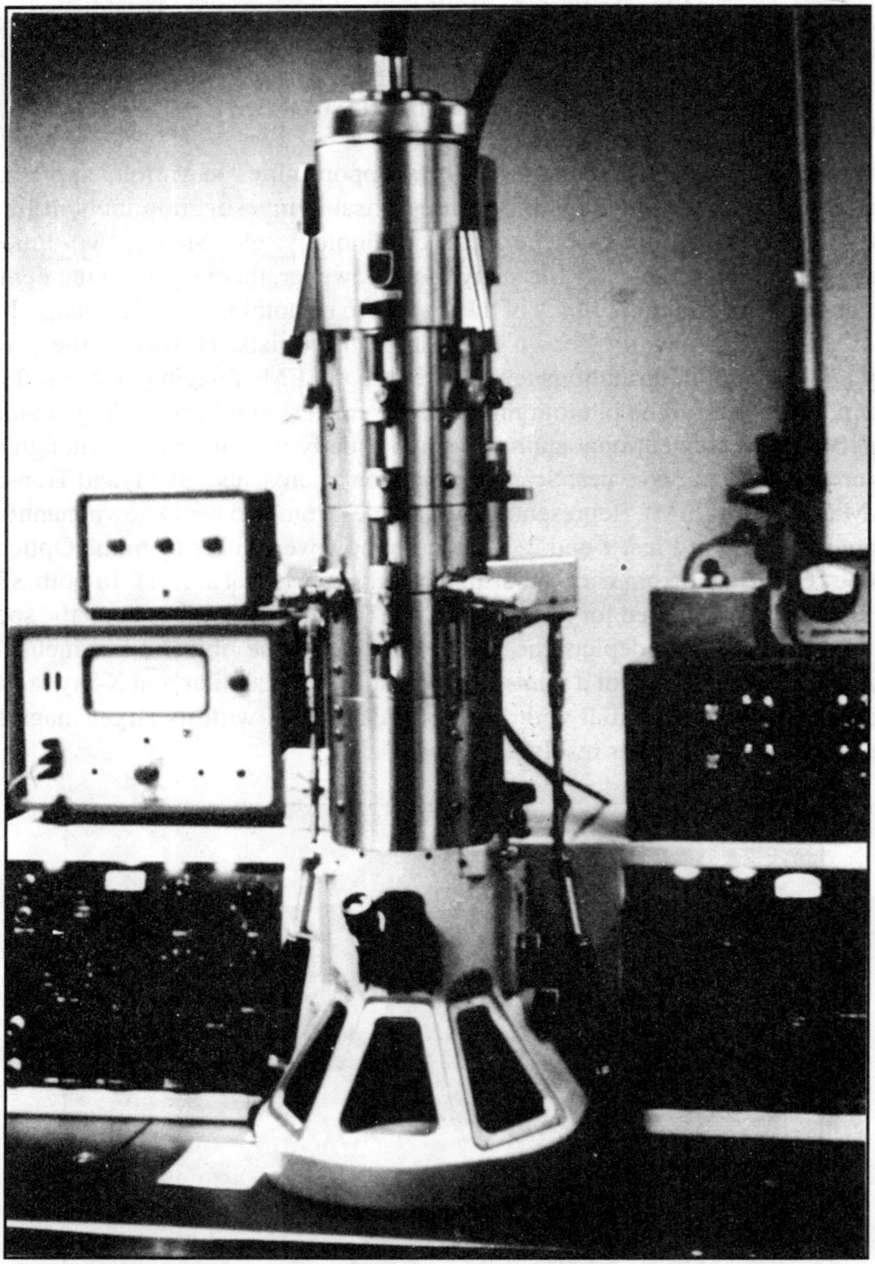

Fig. 2: Philips EM 300, TEM (general view)

Table 1: Comparison of OLM, TEM and SEM characteristics
[After Black, 1970, with updated data]

Characteristics	*OLM*	*TEM*	*SEM*
Lens Systems	Glass or Quartz	Electromagnetic	Electromagnetic
Image Display	Human eye, Film or Frosted glass	Fluorescent viewing plate or film	CRT for viewing and photography
Energy	Radiation from UV to IR (through the visible spectrum)	Thermonic emission of electrons by 50 to 1000 kV	Thermonic emission of electrons by 1 to 30 kV
Wave Lengths (Light: 0.2 μm)	200–700	≥ 0.1 nm	≥ 0.1 nm
Resolution	100 nm	0.3 nm	2.0 nm
Depth of Field	Poor	Good	Excellent
	250 μm at 15x to 0.8 μm at 1200x	500 μm at 4000x to 0.2 μm at 500,000x	1000 μm at 100x to 10 μm at 10,000x
Magnification	15x–3000x	100x–500,000x (Stepped)	10x–1,000,000x (Continuous)
Sample Requirements	Thin sections only	0.1–1 μm thin sections or only replica, 2-3 mm diameter	No thickness limit Non-metals (better) Metal coated (Au, Pd or Al), to 25 mm diameter

1.2.1 Scanning Electron Microscope (SEM)

The SEM, unlike the usual OLM and TEM, is able to image solid objects as it records not the electrons passing through the specimen but the secondary electrons that are released from the atoms of the specimen surface. These help in the formation of the images which appear on its CRT (Fig. 3). The sample can, therefore, be of any size and shape that fits in the evacuated specimen chamber of the SEM.

The imaging system (Fig. 3) uses electron radiation for localization and visible light radiation for information transfer. The resolving power is dictated by electron optical consideration and the information content is dictated by the interaction of visible light in matter. The electron radiation is used for two synchronous scanning beams, viz., (a) a point source of radiation sweeping over the specimen in a well defined pattern of lines with a well defined velocity; and (b) a corresponding second point source of radiation over a fluorescent screen, which is eventually recorded with a camera (usually a polaroid type that gives almost instantaneous pictures). A one-to-one correspondence in time and space exists between the specimen and image points. This, however, is achieved not by optical focussing but by time sequencing.

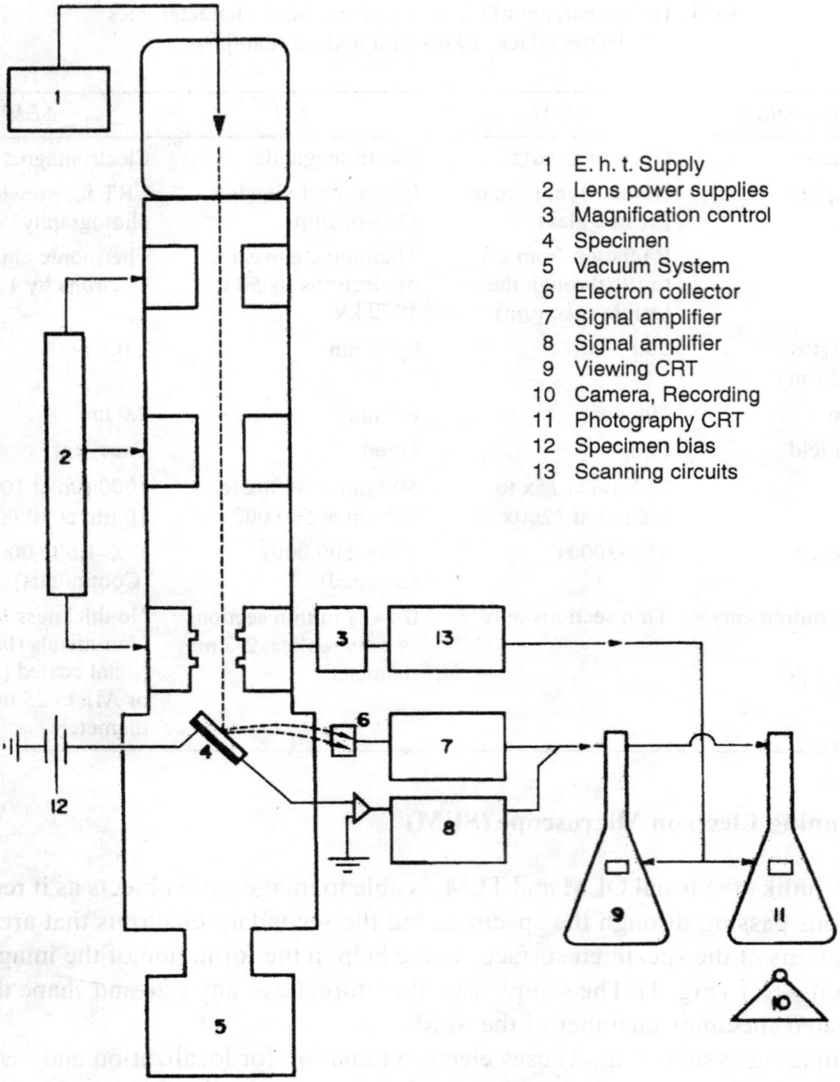

Fig. 3: Schematic diagram of a SEM

The following are the primary factors causing variation in intensity in the SEM images: (i) Those that depend on the local angle of incidence of the electron probe; (ii) Those that cause variations in the fraction of secondary electrons collected by the detector; and (iii) Those that depend on the nature of the material. The first two are similar to the angular relationship of the sun to the topographic features of the earth's surface in aerial or space photography. The third one is analogous to color in an optical image. The importance of these factors is that something can be done to each to enhance the quality of the recorded image. For example, coating of the specimen with selected metallic films can improve the emission characteristics of secondary electrons. Furthermore, the specimen is tilted and the secondary electron collector is so oriented that a majority of secondary electrons are emitted directly towards the collector. Other subtle actions are available to a microscopist who often learns (through experience) how to improve the image quality and to optimize various controls.

Concerning the scanning action and magnification in the SEM is a feature that relates to the size of the picture element (pixel): On the high resolution CRT used for micrography, the minimum spot size diameter is usually 0.1 mm. The corresponding pixel diameter would depend on magnification, e.g., at 10x magnification it is 0.01 mm, while at 100,000x it is 1 nm.

The specimen holder comes in various sizes to accommodate the specimens adequately. Use of specimens upto 89 mm in diameter and 25 mm in thickness have been reported. A micrograph-stereogram of a familiar object (Fig. 4) taken at a SEM would give some idea.

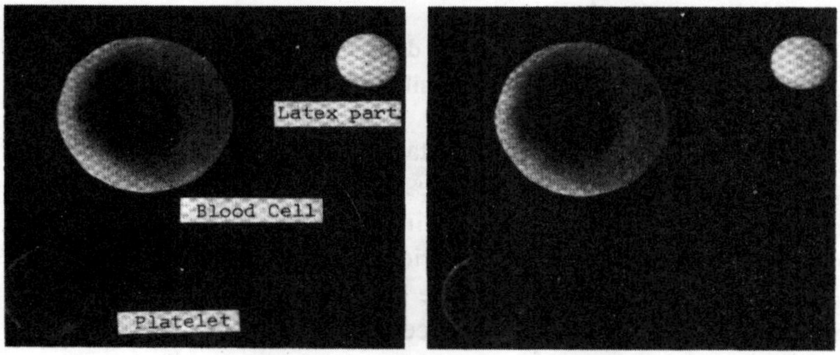

Fig. 4: Stereo-micrograph with a SEM

Showing a human blood-cell and a platelet as well as a latex particle (used for dimensional comparision). Magnification: ~4000x.

Note: The left photo is untilted while the right one has 10° tilt

The specimen stages (single-tilt type) in most SEMs and TEMs have the following possible movements:

(a) *Tilt* (uniaxial, around *X* or *Y* axis), corresponding to ω or ϕ rotation in conventional photogrammetry;
(b) *Rotation* around an axis parallel to the general direction of the electron beam, similar to κ rotation around *Z* axis in conventional photogrammetry; and
(c) *Translatory movements* in *X*, *Y* and *Z* directions.

In a typical single tilt stage in both, SEM and TEM (Fig. 5), the specimen is mounted in a cradle that is trapped on top and bottom in the specimen rod. The cradle is rotated with respect to two pins against which it is trapped by means of a spring-loaded tape. When the tape is pulled, the specimen stage rotation (κ) is effected around the stage normal (*Z* axis). The tilt (ϕ) is around the long axis (*Y*) of the specimen rod. The rotation axis changes orientation with respect to the electron beam axis as a function of the angle ϕ.

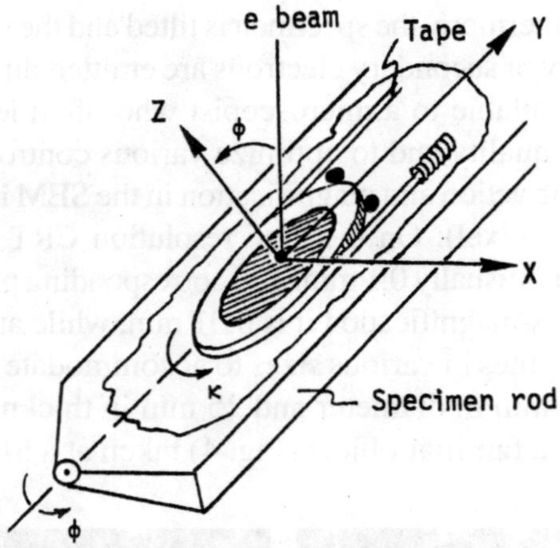

Fig. 5: Typical schematic design for a single-tilt stage (with rotation possibility) in a TEM or SEM

Certain SEMs have double-tilt rotation stages. The tilt angles ϕ and ω are achieved by moving a stage block along arcs (Fig. 6). The *X-Y* translations are located inside the ω and ϕ arcs. Thus, the stage is eucentric. The specimen can also be rotated around an axis parallel to the normal to the stage by either a eucentric or non-eucentric motion. Such single-tilt or double-tilt stages with rotation can provide complete freedom of specimen positioning. The rotation (κ), however, is not always eucentric.

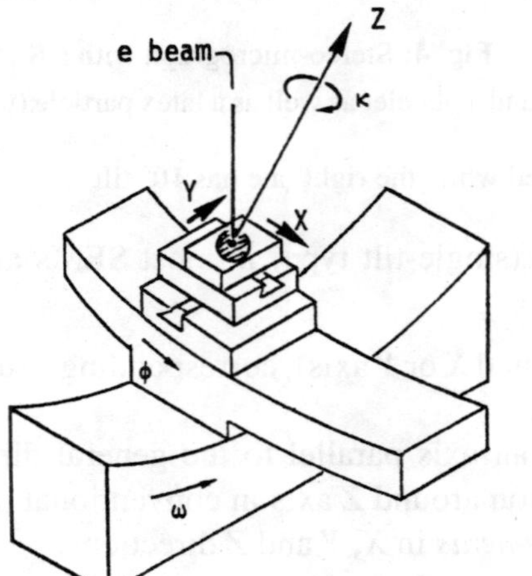

Fig. 6: Typical schematic design for a eucentric double-tilt stage (with rotation possibility) in a SEM

The impinging electron beam on the specimen also dislodges back-scatter electrons, auger electrons, X-rays, light and other detectable effects, which make the SEM an extremely versatile tool. The scanning is done in a raster pattern, as in a standard TV.

The various modes available for image formation provide a wealth of distinctive information about the specimen. The SEM permits observation of such things as conductivity or resistivity in micro-circuits, positive-negative junctions in transistors and magnetic recording patterns on magnetic tape. X-ray analysis of metals and crystals is also an important function of SEMs. Depth of field and magnification range are considerable in SEMs (see Table 1).

1.2.2 Transmission Electron Microscope (TEM)

The TEM is very analogous to the more familiar OLM and is the most common research EM (Fig. 7). Thin sections (usually less than 100 μ m) are mounted on a stage and inserted

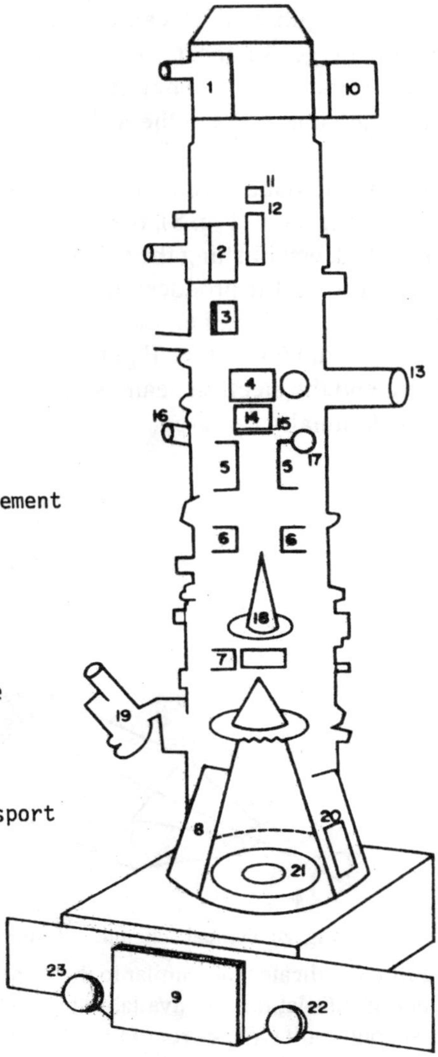

Fig. 7: Schematic diagram of a TEM

into the body of the scope. The scope is then evacuated with a vacuum pump. An electron beam is focussed on the specimen with electromagnetic lenses. The beam passes through the specimen, defocussed below the specimen and projected on to a photo-luminescent screen or photo emulsion. Differential energy losses of electrons, as they are transmitted through the specimen, provide contrast and detail in the projected image.

As in the SEM, there is one-to-one correspondence between the points in the specimen and in the TEM image. There is an electrically controlled shutter (below the projector lens), which permits accurate exposure timing. Further below is a vacuum valve which allows the microscope column to be sealed off from the viewing chamber and camera. Some TEMs use 35 or 70 mm roll film cameras which can be inserted in the viewing chamber while a glass plate camera can be inserted below the viewing screen. The magnification range for most TEMs is from 100x to 500,000x. The resolving powers generally vary from 0.3 to 3.0 nm. However, the image resolution is largely determined by the specimen thickness. As in the SEM, there are also electromagnetic lens aberrations in the TEM.

The specimen stages are of two types (as in the SEM), single and double-tilt. In a typical double-tilt stage (Fig. 8), the specimen is mounted in a cradle which is gimbaled to the specimen rod, establishing the ω rotation axis. The ϕ rotation is around the axis (Y) of the specimen rod and it produces the second tilt to give a stereo-pair. The tilt axes are, however, not independent, i.e., the orientation of the ω axis relative to the electron beam axis changes as a function of ϕ. Figure 8 shows the case where both ω and ϕ angles are not equal to zero and the electron beam is incident on the center of the specimen stage, i.e., the stage is translationally centered.

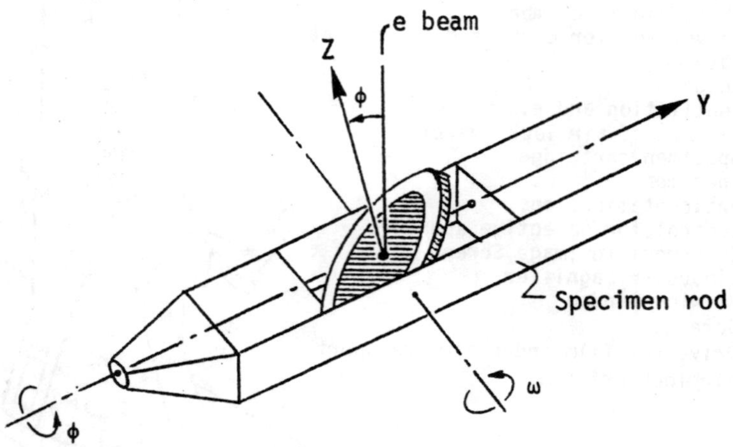

Fig. 8: Typical schematic design for a double-tilt stage in a TEM

The above would indicate that (similar to the SEM) the features giving ω (or ϕ) and κ rotations as also X, Y and Z translations are available in most commercially available TEMs. It may be noted that, in most double-tilt type, κ rotation is absent while both ω and ϕ rotations are available. Such instruments are in much less use, however.

With regard to photogrammetric parameters, one finds a similarity between the two types of EMs. In the metric (quantitative) evaluation of EM micrographs, a major role is played by the resolution capability, which is better than 3 nm in both. This may be the limiting factor for precision measurements under the current circumstances. For example, at a magnification of 5,000x, the resolution of 3 nm would correspond to 0.015 mm on the micrograph. The measuring tool (such as a comparator) having a standard error of less than 0.01 mm on the photo may then be considered adequate. Higher magnifications giving sharper images would demand a measuring instrument of better precision, a more stable emulsion base, etc.

The choice of measuring instrument must also depend on the projection geometry and distortions in the resulting image. In this regard, continuous mapping can be performed where near parallel projection is adaptable and the distortions can be adequately minimized. Analytical plotters with digital mapping capabilities are superbly suited for EM applications. For limited accuracy and for specific applications, analog instruments like Zeiss Stereotope have been found suitable (Boyde and Ross, 1975).

2

GEOMETRIC ASPECTS

An understanding of the various components involved in the EM systems, their geometric forms and their configurations is necessary for precession measurements with micrographs. The implied systems of coordinates, their transformations and consideration of various distortions, although similar in many respects of conventional photogrammetry, are uniquely special in the EMs.

2.1 COORDINATE SYSTEMS

A coordinate system is one of reference for defining points or objects in a specific space by means of distances and/or angles. The reference can be made with regard to designed axes, planes, or surfaces (Ghosh, 2005). In electron microscopy, however, only the *rectangular* systems are useful. They can be either two-dimensional (2-D), where points are defined by linear distances from two mutually perpendicular axes in a plane, or three-dimensional (3-D), where points are located with reference to three mutually perpendicular planes in space. Further, a coordinate system can be used with regard to a photo (micrograph), a stereo-model, or the corresponding object. Accordingly, one would refer to photo-, model-, or object-coordinate systems. These systems can be related to one another through transformation of coordinates (see Figs. 9 and 9a).

A. Photo-Coordinates (x, y)

The measuring of points on a micrograph requires a reference system established with a set of marks (fiducials) which can be observed on the micrograph and which would maintain a fixed relationship to the imaging system (e.g., the recording CRT in a SEM). In the ideal situation, there would be at least four such marks at the extremities of the photo format. An excellent solution would be to have a set of reseau (grid) marks etched on the CRT surface in (SEM) or on a fixed glass plate (in TEM) that could be sharply focussed along with the specimen image by the registering camera. There being no such provision made by the EM manufacturers, it is upto the user to furnish such necessary *fiducial marks.* Ghosh (1975) describes one set of marks created by using epoxy steel fillers on the camera frame

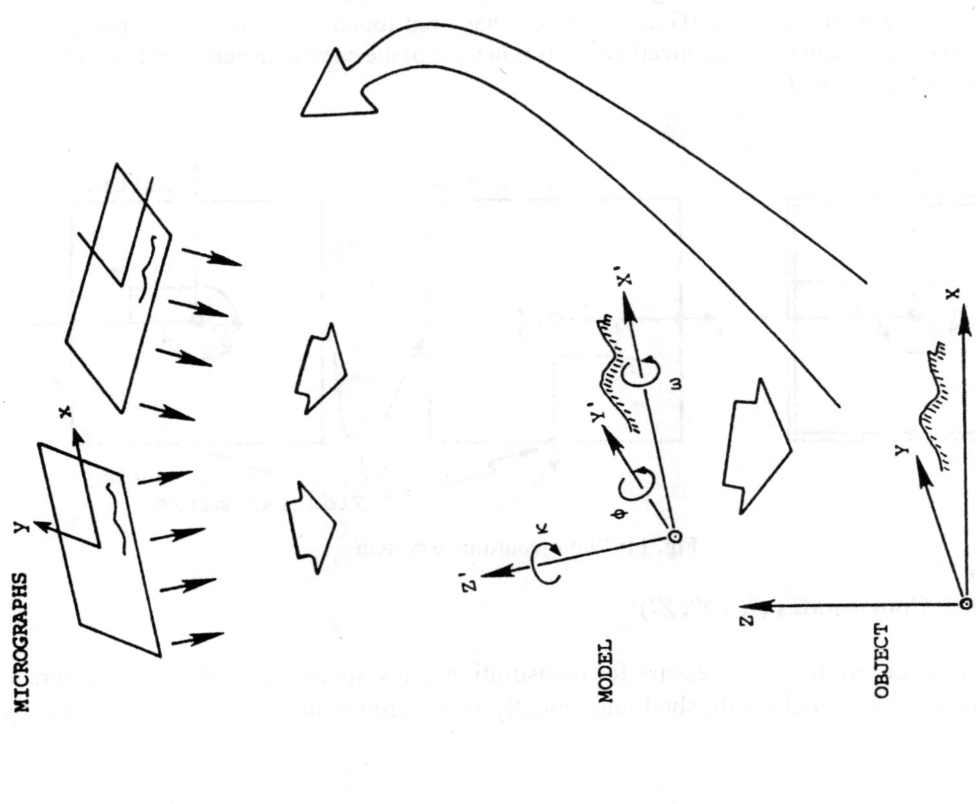

Fig. 9: Systems of coordinates involved in EM applications

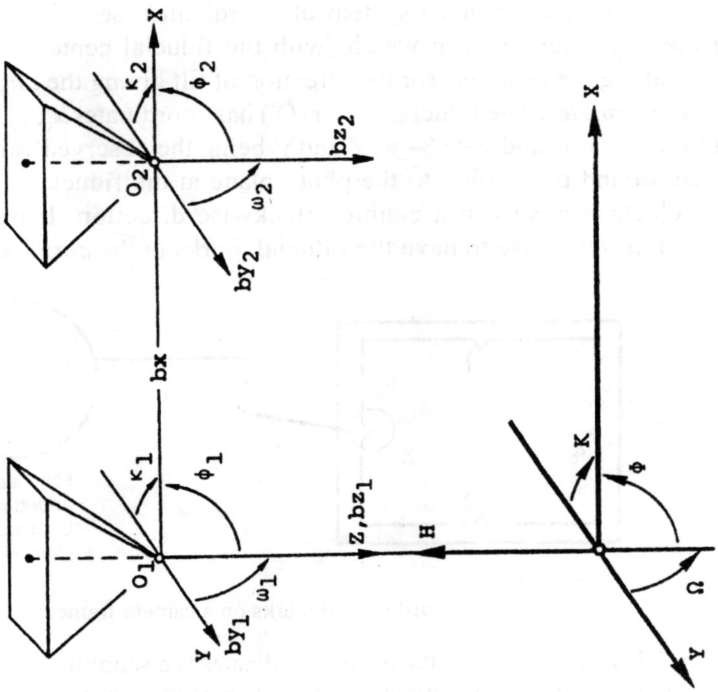

Fig. 9a: Model coordinate system (International)

(see Fig. 10). In a right-handed system of coordinates (see Fig. 11) the fiducial marks provide the *x*, *y* references, in which (with the fiducial center as origin) the x-axis is considered along the *base line* (or the direction of tilt giving the *angle of convergence*, or the *parallactic angle*). The fiducial center (O') has coordinates x_0, y_0. An image point (p) is defined by $x = x' - x_0$ and $y = y_0' - y'$; x' and y' being the observed values. The κ-rotation is considered around the normal to the photo plane at the fiducial center (i.e., around the principal electron beam) in a counter clockwise direction. It may be pertinent (and convenient) in some case to have the fiducial marks in the corners of the image format.

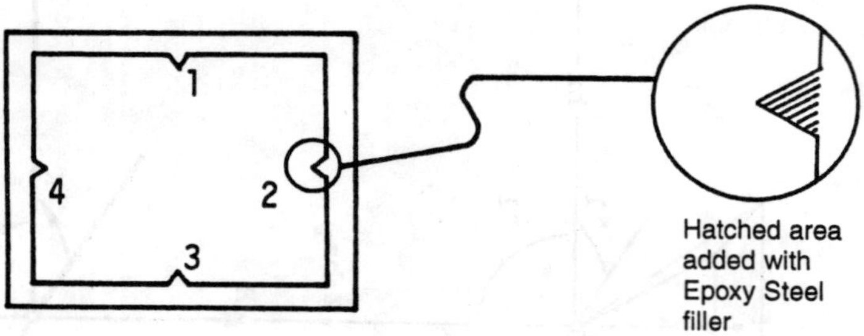

Fig. 10: Fiducial marks on a camera frame

[*Note*: This, 2-D consideration of the photo-coordinates is a simplified form of the 3-D (rigorous) consideration, where the system origin is at the perspective center and a the distance of the *focal length* from the principal point; (Ghosh, 2005). It has been found from experience that this 2-D form is very convenient in EM applications. This is in view of the parallel to near-parallel projection geometry in the EM systems.

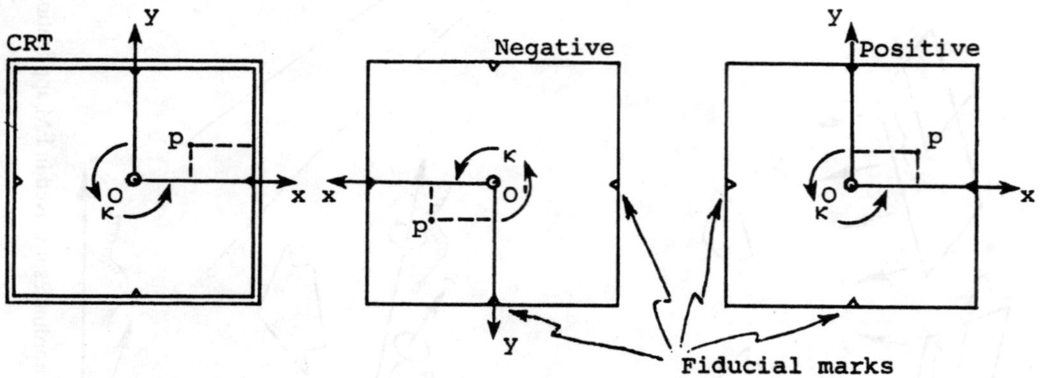

Fig. 11: Photo-coordinate system

B. Model-Coordinates (*X'*, *Y'*, *Z'*)

A 3-D reference system is necessary for mensuration in a solid object. This system refers to the stereoscopic model established analogically at a stereo instrument or mathematically with a computer. This system is considered 3-D, orthogonal and right-handed (Fig. 9a). The orientation of the system in a model and the location of its origin may be arbitrary, depending on the convenience for the use of EM data. In photogrammetric practice, the X'-direction coincides with the base line or the direction of tilt used in creating the stereo model. In usual photogrammetric practice, one prefers to use the *International System* (agreed upon at the International Congress of Photogrammetry in 1960 as in Fig. 9a), in which all model rotations (Ω around X'-axis, Φ around Y'-axis and K around Z'-axis) are positive clockwise when viewed along the positive direction of the particular axis toward the origin. The ϕ rotation corresponds to the angle of tilt used in creating the stereo pair. The model may be considered as a stage intermediate between the object and the corresponding micrographs, and this is where the mensural (quantitative) data are obtained. Finally, the model data need to be related to the object with due consideration of the scale factor.

C. Object-Coordinates (*X*, *Y*, *Z*)

Finally, this is the 3-D system utilized for defining the position of points in the object space, as distinguished from the image or the model. It is also considered right-handed and orthogonal. Depending on the specific application, the choice of the origin or the orientation of the system can be arbitrary. The required depiction of the object in terms of applications like mapping or modelling would be logically made in this system (See Fig. 9).

2.2 TRANSFORMATION OF COORDINATES

It is often necessary to establish relationship between coordinate systems for relating points of one system to the other. This is performed through transformation of coordinates. There are various methods of transformation for quantitative applications (Ghosh, 2005). This may mean a simple change in location and orientation (i.e., with no change in shape and size) or a complex one (with change in both size and shape) or even something in between, e.g., variation in size (scale) with no change in shape or location or attitude (orientation).

A. Similarity Transformation

For most EM work, a simple 3-D (*similarity*) transformation is very effective. Basically, it considers only changes in location, orientation and size (scale), yet it is adaptable to considerations of minor changes in shape. It is symbolized as follows:

$$X = k\, M^T\, \bar{X} + X_O \qquad \text{Eq. (1)}$$

where

$X = [X\,Y\,Z]^T$ coordinates after transformation;

$\bar{X} = \left[\bar{X}\,\bar{Y}\,\bar{Z}\right]^T$ coordinates before transformation;

X_O = $[X_O\, Y_O\, Z_O]^T$ vector of three shifts indicating the X, Y, Z coordinates of the orign of the $\bar{X}, \bar{Y}, \bar{Z}$ system;

M = an orthogonal three-angle rotation matrix, consisting of three sequential rotations;

and

k = a scale factor.

Considering the rotations sequentially, ω around $\bar{X}$ (primary rotation), ϕ around $\bar{Y}$ (secondary rotation) and k around $\bar{Z}$ (tertitary rotation), their effects are expressed by

$$M = M_\kappa \,.\, M_\phi \,.\, M_\omega \qquad \text{Eq. (2)}$$

The elements of this rotation matrix can be expressed in various ways as explained in Ghosh (2005).

The general transformation expressed by Eq. (1) assumes that the scale is the same in all directions. It has, however, been noticed by many in EM applications that there may exist different scales along different directions in a micrograph or in a stereo-model. Such a condition causes geometric *affinity* and corresponding variation in the shape of the object as registered on the micrograph or as measured in the stereo model. The problem of such cases can mostly be alleviated in most applications with Eq. (1) being modified by using $k = 1$ and

$$\bar{X} = \left[k_{\bar{X}} \bar{X}\; k_{\bar{Y}} \bar{Y}\; k_{\bar{Z}} \bar{Z}\right]^T \qquad \text{Eq. (3)}$$

where $k_{\bar{X}}$, $k_{\bar{Y}}$ and $k_{\bar{Z}}$ are the respective scale factors along $\bar{X}, \bar{Y}$ and $\bar{Z}$ axes.

In the utilization of a transformation, it is often found convenient to proceed in a sequential pattern with intermediate steps as illustrated in Fig. 12 with regard to establishing a relationship between the micrograph and the corresponding object coordinates. These steps can be performed also in the reverse mode, i.e., from the object to micrographs.

B. Polynomial Transformation

Sometimes (rarely in EM applications), in view of propagated errors or systematic deformations in the micrographs, a stereo model or a series of models in a row (to make a complete coverage of the sample/object), one may use polynomials of various forms (Ghosh, 2005). In reality, such polynomials may be composed of non-linear, trigonometric or other functions. Such polynomials, therefore, are not conformal and should be used only in case where the rotation angles are small. The use of such polynomials may not be able to provide information on actual physical parameters involved in the system. They can be, nonetheless, useful in performing the necessary numerical transformations, as well as in evaluating the quality of the utilized data.

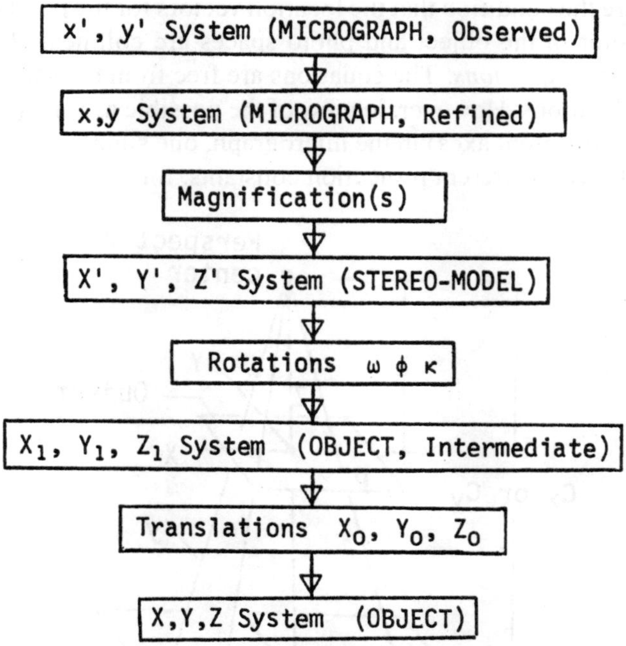

Fig. 12: Scheme of typical sequential steps in transformation

2.3 PROJECTIONS AND DISTORTIONS

A photograph may be considered as a projected record of information. Also in conventional photogrammetry it is considered to be 3-D (*x, y, z),* where the third dimension (*z*) is a constant (-C, known as *camara constant* or *focal length)*. By using Eq. (1), one can establish the projective relationship between the photo and the object through three equations as represented in Eq. (4), where the m's are the elements of the *rotation matrix* (M):

$$\begin{bmatrix} x \\ y \\ -C \end{bmatrix} = k \begin{bmatrix} m_{11} & m_{12} & m_{13} \\ m_{21} & m_{22} & m_{23} \\ m_{31} & m_{32} & m_{33} \end{bmatrix} \begin{bmatrix} X - X_O \\ Y - Y_O \\ Z - Z_O \end{bmatrix} \qquad \text{Eq. (4)}$$

If the resulting first and second equations are divided by the third, one obtains the following equations:

$$x = -C\,\frac{(X - X_O)m_{11} + (Y - Y_O)m_{12} + (Z - Z_O)m_{13}}{(X - X_O)m_{31} + (Y - Y_O)m_{32} + (Z - Z_O)m_{33}}$$

$$y = -C\,\frac{(X - X_O)m_{21} + (Y - Y_O)m_{22} + (Z - Z_O)m_{23}}{(X - X_O)m_{31} + (Y - Y_O)m_{32} + (Z - Z_O)m_{33}} \qquad \text{Eq. (5)}$$

Equations (5) ensure the condition that the location vectors for the point (P) with respect to the perspective center in the object and photo spaces are collinear (Fig. 13). These are known as *Collinearity Equations.* The equations are free from the scale factor (k) and are useful in many applications. However, in view of the possible affinity (i.e., different scale/magnification along different axes) in the micrograph, one simple way of modifying them would be to consider two different projection constants, for the two equations.

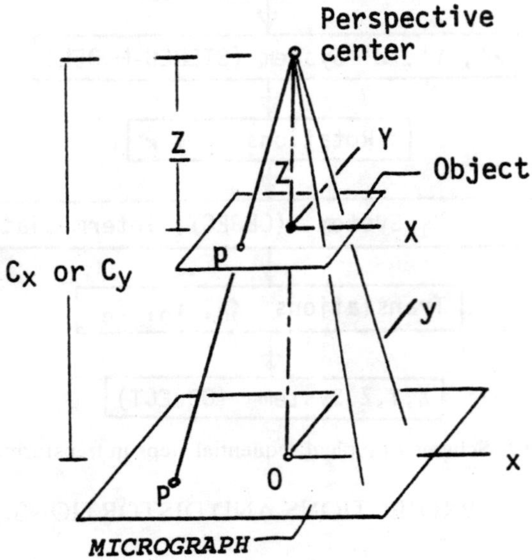

Fig. 13: Perspective projection and collinearity condition

A. Parallel Projection

In view of extremely small objects and extremely large magnifications, an electron micrograph has an extremely small field angle and the projeced bundle of rays can be viewed as parallel or near-parallel projection (Fig. 14), the magnification being a sequence before or after the projection. The parallel projection can be expressed by:

$$\begin{bmatrix} x \\ y \\ 0 \end{bmatrix} = \begin{bmatrix} K_1 & 0 & 0 \\ 0 & K_2 & 0 \\ 0 & 0 & 0 \end{bmatrix} \begin{bmatrix} X_R \\ Y_R \\ Z_R \end{bmatrix} \qquad \text{Eq. (6)}$$

where x, y are the photo coordinate;

K_1, K_2 are the scale factors along Y_R, Y_R directions, respectively;

and X_R, Y_R, Z_R are the object-space coordinates (considering rotations).

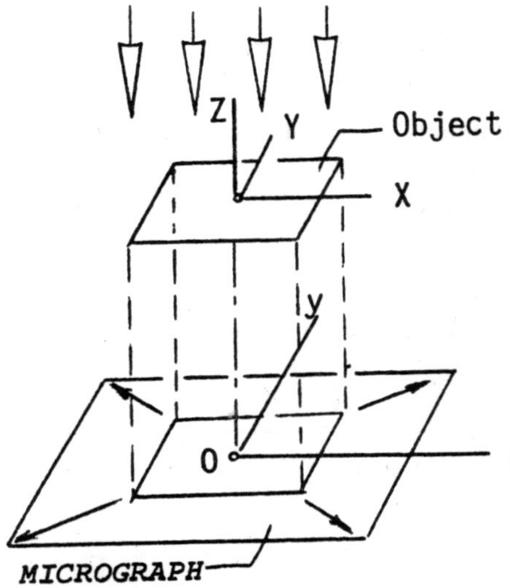

Fig. 14: Parallel projection and magnification

In view of rotations and translations to a selected origin the last term in Eq. (6) becomes:

$$\begin{bmatrix} X_R \\ Y_R \\ Z_R \end{bmatrix} = M \begin{bmatrix} X - X_O \\ Y - Y_O \\ Z - Z_O \end{bmatrix} \qquad \text{Eq. (7)}$$

where X_O, Y_O, Z_O are the coordinates of the selected origin in the object systems; and M is the orientation matrix.

If the height datum (level reference) is defined by setting $Z_O = 0$, after substitutions and simplifications (Ghosh, 2005), one obtains the following projective equations in the case of parallel projection:

$$x = K_1(X-X_O)(c\phi\ c\kappa + s\omega\ s\phi\ s\kappa) + K_1(Y-Y_O)c\omega\ s\kappa + K_1 Z(s\omega\ c\phi\ s\kappa - s\phi\ c\kappa)$$
$$y = K_2(X-X_O)(s\omega\ s\phi\ s\kappa - c\phi\ s\kappa) + K_2(Y-Y_O)c\omega\ c\kappa + K_2 Z(s\phi\ s\kappa + s\omega\ c\phi\ c\kappa) \qquad \text{Eq. (8)}$$

where *c* and *s* prefixes mean cosine and sine functions, respectively, of the rotation angles as indicated.

B. Perspective Projection

There are two principal features of a perspective projection : (a) A *perspective center* from where the rays appear to diverge, and (b) A finite *projection distance* (Fig. 13). In view of the previous discussions, the following relations hold:

$$x = X_{\mathrm{R}} \frac{C_{\mathrm{x}}}{\underline{Z} - Z_{\mathrm{R}}} \text{ and } y = Y_{\mathrm{R}} \frac{C_{\mathrm{y}}}{\underline{Z} - Z_{\mathrm{R}}} \qquad \text{Eq. (9)}$$

where $\underline{Z}$ is the projection distance to the reference dadum of the object, and the rest of the terms are as defined previously.

By substituting the expansion for Z_R from Eq. (7) into Eq. (9), after rearranging,

$$x = \frac{(C_{\mathrm{x}} / \underline{Z}) X_{\mathrm{R}}}{1 - \frac{1}{\underline{Z}}\{(X - X_{\mathrm{O}}) c\omega c\phi - (Y - Y_{\mathrm{O}}) s\omega + (Z - Z_{\mathrm{O}}) c\omega c\phi\}}$$

$$y = \frac{(C_{\mathrm{y}} / \underline{Z}) Y_{\mathrm{R}}}{1 - \frac{1}{\underline{Z}}\{(X - X_{\mathrm{O}}) c\omega c\phi - (Y - Y_{\mathrm{O}}) s\omega + (Z - Z_{\mathrm{O}}) c\omega c\phi\}} \qquad \text{Eq. (10)}$$

One can substitude K_1 for $C_{\mathrm{x}} / \underline{Z}$ and K_2 for $C_{\mathrm{y}} / \underline{Z}$. Furthermore, the second term in the denominator being smaller than unity, the expression can be expanded in a series (cf. $1/\{1 - q\} = 1 + q + q^2 + q^3 + \ldots$).

This gives the final expressions for a perspective projection:

$$x = K_1 X_{\mathrm{R}} \{1 + \nabla + \nabla^2 + \ldots\}$$
$$y = K_2 Y_{\mathrm{R}} \{1 + \nabla + \nabla^2 + \ldots\} \qquad \text{Eq. (11)}$$

where ∇ is the second term in the denominators of Eq. (10). Details of the derivation and the validity of various approximations are credited to Nagaraja (1974).

C. Distortions

With regard to EMs, it is practical and convenient to consider the parallel projection as appropriate. The deviations from the parallel projection may be considered as *distortions*. The various distortions and the corresponding mathematical models are given below in accordance with their relative magnitudes (as established from research studies, see Adiguzel, 1985; El Ghazali, 1978; Ghosh and Nagaraja, 1976 and Maune, 1973). See Fig. 15 for their patterns:

(1) Perspective Distortion

This defines the difference between the perspective and parallel projections. One may note that the first terms of Eq. (11) constitute Eq. (6). This gives:

$$x_{\text{parallel}} = x_{\text{pers}} - x_{\text{pers}} \{\nabla + \nabla^2 + \ldots\}$$

$$y_{\text{parallel}} = y_{\text{pers}} - y_{\text{pers}} \{\nabla + \nabla^2 + \ldots\} \qquad \text{Eq. (12)}$$

The representative values of ∇, ∇^2 etc. can be obtained from calibration studies of the specific EM imaging system.

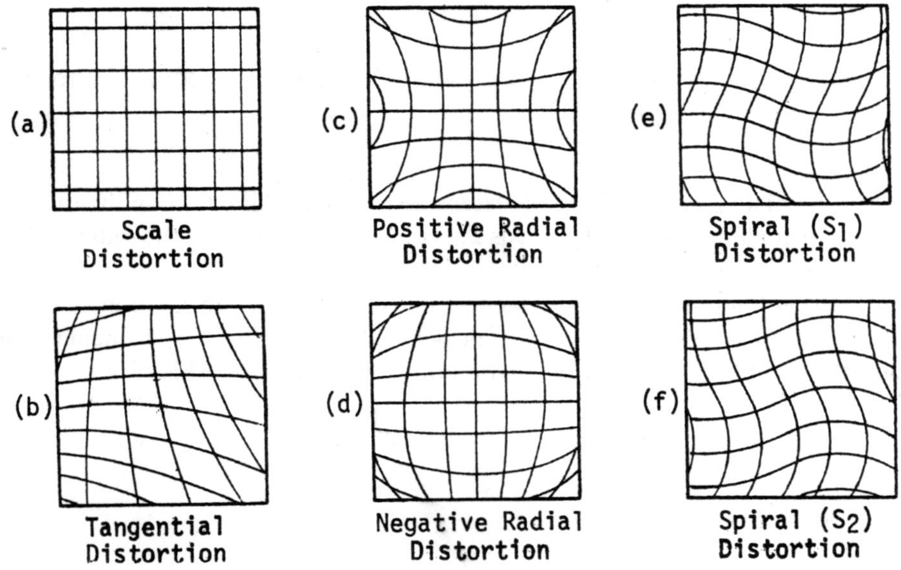

Fig. 15: Distortion patterns in EM

(2) Scale Distortion

Scale distortion or affinity (in view of different scales in different directions) is contained in Eqs. 6, 9, 10 or 11 and is, therefore, best dealt with directly in the mathematical model for the projection.

(3) Radial Distortion

This is considered, conventinally, with regart to the fiducial center/principal point of the micrograph, either positive (in the outward direction) or negative (in the inward direction). similar to conventional photogrammetry, it is expressed by a polynomial:

$$\Delta\ r = k_0 r + k_1 r^3 + k_2 r^5 + \dots \qquad \text{Eq. (13)}$$

where r is the radial distance of the image point from the photo center; and k's are certain constants. The first term (k_0), however, is equivalent to a scale factor and in a sequential application of corrections, can be considered as contained in the terms K_1 or K_2.

It has been found from actual studies that terms of 5 and higher order in r can be neglected in normal applications. For the x and y components, therefore, one gets;

$$\Delta x = \Delta r \frac{x}{r} = k_1 \frac{r^3 x}{r} = k_1(x^3 + xy^2) = D_1 x^3 + D_2 xy^2$$

$$\Delta y = \Delta r \frac{y}{r} = k_1 \frac{r^3 y}{r} = k_1(x^2 y + y^3) = D_3 y^3 + D_4 x^2 y \qquad \text{Eq. (14)}$$

where the D′s are certain constants.

(4) Spiral Distortion

Klemperer and Barnett (1971), expressed the spiral or rotational twist of the electron beam by

$$\Delta\ d = C(y_0'/y)r^3 \qquad \text{Eq. (15)}$$

where $\Delta\ d$ is the displacement of the ray lateral to the radial direction on the micrograph; C is the spiral distortion coefficient; y_0'/y is the magnification and r is the radial distance of the point from the principal electron axis.

Magnification being already contained in the K factors, a substitution, $S = C(y_0'/y)$, may be considered. Furthermore, in view of possible affinity, it may be pertinent to consider two values of S: S_1 and S_2 for the x and y directions, respectively. This gives:

$$\Delta x = S_1 \frac{y}{r} r^3 = S_1(x^2 y + y^3)$$

$$\Delta y = S_2 \frac{x}{r} r^3 = S_2(x^3 + xy^2) \qquad \text{Eq. (16)}$$

where S_1 and S_2 are the spiral constants.

This distortion pattern should not be confused with the striking phenomenon of systematic rotation of the entire beam which happens on changing the magnification (Ghosh and Adiguzel, 1984). Such rotations can be of rather large magnitudes (Fig. 16). However, as long as the same magnification is used for data acquisition, such a systematic rotation, having been contained in transformation, would have no adverse effect on mensuration.

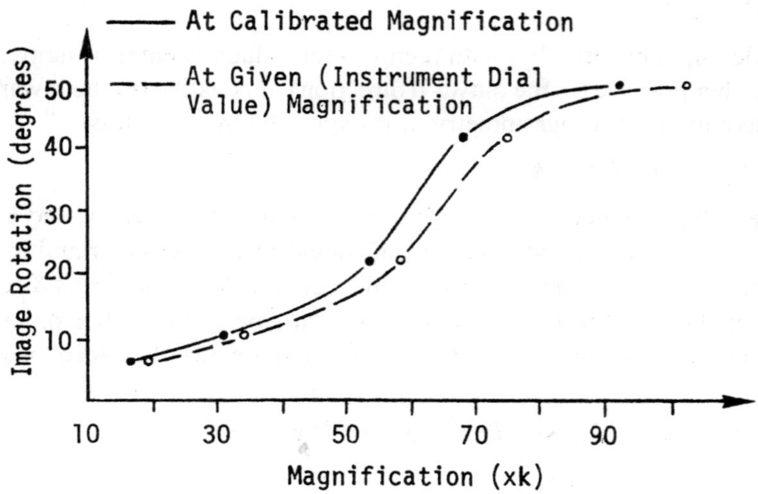

Fig. 16: Image rotation for magnification in one TEM [Example from Ghosh and El Ghazali, 1977]

(5) Tangential Distortions

Maune (1973) modeled this pattern of distortion after that of a conventional photogrammetric camera (effect due to decentering of lens) and obtained satisfactory results. However, Nagaraja (1974) found that this type of distortion can be effectively accommodated in the mathematical model for spiral distortion. Adiguzel (1985) found some (although not very significant) improvement in the mensural accuracy by considering this distortion. Since the effect of tangential distortion is small and often very negligible, and the methematical model is somewhat lengthy, it is not presented here. The interested reader may obtain ideas in Ghosh (2005) or the ASPRS Manual (2004).

The degree of *stability* of an EM system has a significant influence on the distortions. The reliability of the mathematical models, therefore, would depend on the calibration and evaluation of the specific imaging system performed under specific working conditions. The degree of refinement (and confidence) attainable by considering such distortions is illustrated in Fig. 17.

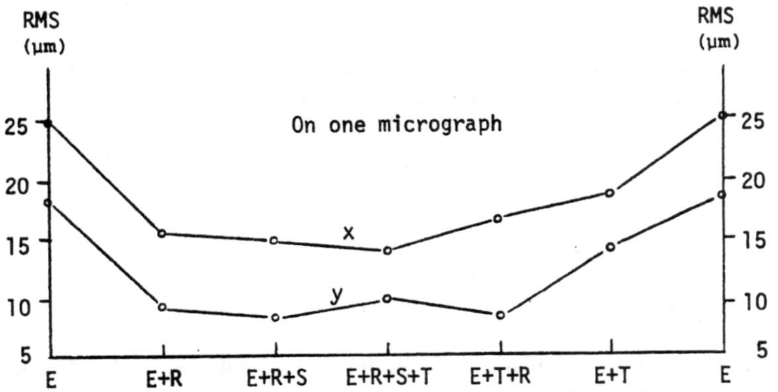

Fig. 17: Standard (RMS) errors of image coordinates with various combinations of correction parameters.

[E = Exterior orientation including magnification, R = Radial distortion, S = Spiral distortion, T = Tangential distortion] (Example from Adiguzel, 1985)

2.4 STEREO MODEL AND ORIENTATION

As in conventional photogrammetry, a 3-D stereo model (replica) of the object can be formed by means of intersection of conjugate rays established by using two micrographs (photos) of the same object. In conventional photogrammetry, the location and directional attitude of the photo or camera (containing the projected bundle of rays) are defined by "exterior" orientation elements consisting of translations along and rotations around the *X*, *Y* and *Z* axes. In EM photogrammetry, however, there is no "exterior" space and the elements of orientation obtained with regard to the specimen (stage or plate) are of prime importance. The various possible movements (orientation elements) in an EM system are discussed in Chapter 1. However, there are limitations in each of these elements, peculiar to each imaging system.

The spatial intersection of conjugate rays is performed by the procedure of *Relative Orientation** (Ghosh, 1972 and Ghosh, 2005), which ensures the condition of *coplanarity*, i.e., all pairs of conjugate rays to be coplanar with the base. Relative orientation with EM photos becomes extremely simple if, for generating the second photo, one uses only the ϕ -tilt and the associated *x*-translation. Subsequent to the relative orientation, one would require the final operation, *Absolute Orientation**, before any meaningful 3-D data can be extracted. Absolute orientation involves *Scaling* and *Levelling* (of the relatively oriented model) with respect to the *Control* data. See discussion on *control* later in this section. The orientations and data acquisition are facilitated if the first micrograph is taken with a tilt angle amounting to half of the *parallactic angle* ($\theta = \gamma /2$ in Fig. 18), inwards, and the second micrograph is taken with the same tilt angle in the opposite direction (also inwards, giving $\theta' = -\theta$). The pre-tilt (θ), if any, would then contribute to tilting the stereo model, i.e., *Levelling* of the elevation (height) datum as necessary in the absolute orientation.

**Relative Orientation*

The general expression (correction equation) for an *analogial (empirical) relative orientation* [mostly used in practice, including at analytical plotters and systems] is:

$$-Py = \frac{XY}{Z}(d\phi_{\text{I}} - d\phi_{\text{II}}) - \left(Z + \frac{Y^2}{Z}\right)(d\omega_{\text{I}} - d\omega_{\text{II}})$$

$$-\frac{Y}{Z}(dbz_{\text{I}} - dbz_{\text{II}} - b \cdot d\phi_{\text{II}}) + X(d\kappa_{\text{I}} - d\kappa_{\text{II}})$$

$$+(dby_{\text{I}} - dby_{\text{II}} + b \cdot d\kappa_{\text{II}}).$$

Here, *Py* refers to the *Y*-parallax (separation in model *Y* coordinates with regard to the two overlapping images when the model *X* coordinate is the same); *X*, *Y* and *Z* are the 3-D model coordinates (giving the point location); κ , ϕ and ω are the rotation elements; by and *bz* are the *y* and *z* translatory elements; *b* is the model base; and the subscripts I and II refer to the left and right side camera (photo carrier), respectively.

Analytically, the coplanirity equation in vectorial notation is: $\vec{b} \cdot \vec{R}_1 \times \vec{R}_2 = 0$. This equation simply states that the three vectors, $\vec{b}$, $\vec{R}_1$ and $\vec{R}_2$ [representing the model base, and the two location vectors for the rays of the points from the two projection (perspective) centers] all lie in the same plane. For details refer to Ghosh (1972, 2005) or ASPRS (2004).

**Absolute Orientation*

The general expression for an absolute orientation [used in practice, which, in a sense, is a 3-D transformation of the model into the system used as control] is:

$$\begin{bmatrix} X^g \\ Y^g \\ H^g \end{bmatrix} = k[M]' \begin{bmatrix} X \\ Y \\ H \end{bmatrix} + \begin{bmatrix} X_0 \\ Y_0 \\ H_0 \end{bmatrix}$$

(Contd.)

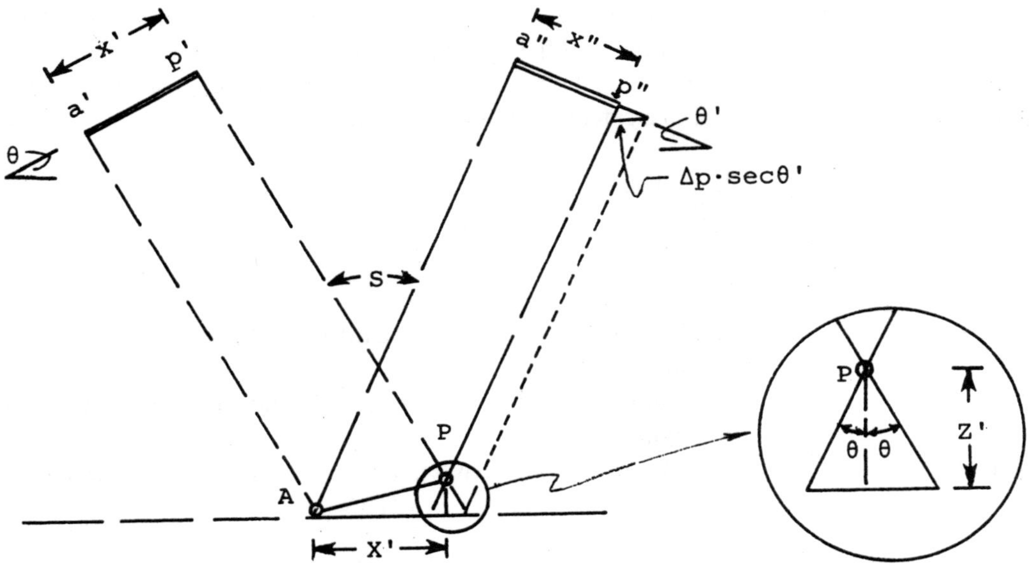

Fig. 18: Intersection geometry, assuming parallel projection

With regard to numerous possibilities and procedural steps in relative and absolute orientations, extensive ideas may be obtained from existing publications in this subject area (e.g., Ghosh, 1972; Ghosh, 2005 and ASPRS Manual of Photogrammetry, 2004). For relative orientation, depending on the number of elements to be solved for, five or less

---

(Contd.)

Here, the superscript *g* refers to ground (control) system of coordinates; *X*, *Y* and *Z* (without any superscript) are the 3-D model coordinates (giving the point location); and the rest as follows:

k	is the scale factor
$[M]'$	is the transposed orientation matrix
$X_0Y_0H_0$	are the coordinates of the origin (0) in the *XYH* (-*Z*) system.

This origin (0) may be any one of the following:

(a) one of the camera stations,
(b) one of the control points, or
(c) center of the area being considered.

The last one, (c) is generally considered in practice.

The scale factor *k* being constant, in practice, is automatically applied along with the elements of the orientation matrix *M*.

The rotation angles, considered sequentially, are:

$$\Omega = \arctan(\Delta H/\Delta Y)$$
$$\Phi = \arctan(\Delta H/\Delta X)$$
$$K = \arctan(\Delta X/\Delta Y)$$

number of stereo observations in a model would be adequate. For absolute orientation, three non-collinear "control" points (with 3-D data) in the model would be adequate. One obtains finally the 3-D coordinates in the oriented model. With parallel projection, the 3-D data of a point P with regard to a reference point (B in Fig. 18) are:

$$Z' = \Delta p \cdot \frac{1}{2\sin\theta} = (x' - x'')\frac{1}{2}\ \text{cosec}\ \frac{\gamma}{2}$$

$$X' = x'\sec\theta\ - Z'\tan\theta \qquad \text{Eq. (17)}$$

and $[Y' = y' = y''$

where x', y' are the photo coordinates with regard to the reference *B* in the first (left) photo;

x'', y'' are the photo coordinates with regard to the same reference in the second (right) photo;

and $\Delta\ p = x' - x''$ is the parallax difference between the observed point *P* and the reference *B*.

Considering affinity, i.e., different magnifications, M_x and M_y in *x* and *y* directions, respectively, on the micrographs, Eq. (17) can be expressed as follows:

$$\begin{bmatrix} X' \\ Y' \\ Z' \end{bmatrix} = \begin{bmatrix} \sec\theta / M_x & 0 & 0 \\ 0 & 1/M_y & 0 \\ 0 & 0 & \text{cosec}\theta / 2M_x \end{bmatrix} \begin{bmatrix} x' \\ y' \\ \Delta p \end{bmatrix} \qquad \text{Eq. (18)}$$

The final data in the object space is then subject to simple transformation from the *X'*, *Y'*, *Z'* (Model) system into the *X*, *Y*, *Z* (Object) system. However, this step may not be necessary at all in most cases.

For the sake of completeness, *Interior orientation* may be mentioned here. This, in conventional photogrammetry, deals with the recovering of the projected cone (bundle) of rays to be geometrically identical with the cone of rays that entered the camera lens to make the original exposure. In EM applications, there being no "camera focal length", the only consideration in this regard is to place the micrograph with regard to the *Fiducial marks* in order to establish the photo-coordinates (Sec. 2.1) as necessary.

With regard to the accuracy of mensural data, one must consider several points:

1. *Stability* of the imaging and the measuring equipment systems. All studies made so far indicate that most modern EM systems are stable for practically all applications. Scale repeatibility (as assessed by measuring the same distance in multiple micrographs taken under the same operational conditions and settings) indicated as much as ±1.17% for one SEM and ±2.07% for one TEM as the average in one study (Ghosh *et al.*, 1978). Such values must be obtained empirically.

2. *Resolution capacity* of the EM imaging system against *Mensuration Capability* of the measuring instrument. Empirical studies in this regard would indicate the optimum capacity of the total system. Table 2 with data from El Ghazali (1978) would illustrate the point.

Table 2: Pointing accuracies (static positional) on Electron Micrographs at grid intersections [*courtesy*, El Ghazali, 1978]

Magnification (x k) [calibrated value]	*Instrument*	*Standard accuracy nm*
6.2	SEM + PSK*	± 4.2
11.1	SEM + PSK	± 3.6
17.0	SEM + PSK	± 7.4
17.0	TEM + PSK	± 2.0
52.6	TEM + PSK	± 2.5
91.0	TEM + PSK	± 3.3

Note: * PSK is the precision stereocomparator by Zeiss.

It may also be noted that the pointing accuracy is a function of the type of point observed (involving its image quality). The data in Table 2 are based on reading grid intersection points where the lines (due to extreme magnifications) are not always sharp. The same comparator (with least count of 1μ m) when used in locating centers of carbon-black spheroids in 18k TEM micrographs gave a standard pointing accuracy between a maximum of ±0.7nm and a minimum of ±0.2 nm.

3. The *Parallactic Angle* would influence the quality of intersection geometry. Theoretically, the best intersection is when the angle is 90°. However, very large angles may create stereoscopic gaps and other related mapping problems. Very small intersection angles would give unreliable heights (elevations). For a given object, a practical compromise between the two extremes is therefore necessary. Ghosh and El Ghazali (1977) reported a parallactic angle between 8° and 20° would generally give adequate results, the optimum being between 10° and 15°.
4. Provision of *Control* points, as necessary for absolute orientation in conventional sense is impossible in EM applications. On the other hand, problems concerning multiple models (strips or blocks) not being necessary, certain innovative approaches in scaling and levelling the model would be very effective. On should, however, use a “standard” for dimensional control.

Scaling (or scale correction) of the stereo model can be performed by making measurements against dimensions of known values. Ghosh (1971) used latex particles as “standards” in mapping blood cells and platelets (Fig. 4) with SEM micrographs. The best available inexpensive standards are carbon-replicas made from master diffraction gratings for TEMs and metal-coated calibration specimens for SEMs. These are cross-ruled grids of fine grooves in a plane surface. The replica grids often used for high magnification give 2160 lines per mm (e.g., manufactured by SIRA [Scientific and Industrial Research Association] or by Ernest F. Fullam, Inc.). The line frequencies are generally guaranteed by the manufacturer to be within 1 per cent of the stated figure. The grids are 2-D only. However, certain configurations of convergent photography can be used to generate the third dimension (see later, discussions on calibration) and thus the stereo-model can be fully controlled in 3-D for assuring precision measurements.

Levelling of the model, being generally arbitrary and related to an assumed reference datum plane as decided by the user, control is no problem.

Note: All modern photogrammetric analytical plotters (or systems) are duly programmed for all orientation operations (interior, relative and absolute).

2.5 CALIBRATION OF AN EM SYSTEM

Calibration is a refined form of measurement for the purpose of evaluating the performance of the working system. An EM system is essentially a man-material-equipment-technique combination. Calibration is better done by assigning numbers to specific elements or parameters used in the process with statistically sound expressions for their systematic errors and precisions. The involved scientist-engineer needs to first establish specifications indicating permissible ranges (or variations) and, next, a state of statistical control. There are, however, certain difficulties, which must be considered:

- The working system may be sensitive to factors other than the desired input (i.e., the object and the micrograph). It is essential that the errors contributed by such factors (such as room temperature or material used for the micrograph) are adequately known, so that they can be eliminated or at least reduced sufficiently. Thus it is necessary to calibrate the system under circumstances and working conditions under which the calibration results will be applied.
- There should be acceptable *Standards* against which both the input and output can be judged. Such standards may not be available for direct use and are often assumed or generated. Their dependability would then rest on the guidelines set up for the purpose. It is desirable that one uses reference standards of proven stability with acceptable repeatability in the working system.
- The working system may not be adequately "stable" and may fail to give the same output for repeated applications. This implies that the results of calibration should remain within a permissible range over the time period for which such results are applied. Ideas on this are already given in Sec. 2.4. Two assumptions are made: that the factors causing the outputs to be in error are randomly active, and that they are not correlated to one another. With multiple observations, the difference of the individual values from the mean (or accepted as correct) value would help one to calculate the standard deviation of the observation data. There is, however, a constant difference between the input and the mean value of the output, known as

bias. Furthermore, *precision*, indicating the 'scattering' of the output values measure the ability of the working system to give the same output for repeated applications at a given input. The *standard deviation* would indicated the *accuracy* with which the outputs can be related to the input.

With regard to EM systems, the following parameters are essential for all mensuration related calibrations:

(1) The locations of fiducial marks, relative to each other, on the micrograph;
(2) The magnification with the consideration of affinity, if known to exist;
(3) All distortion parameters in consideration of the mathematical models to define the systematic deviations of points from their ideal locations on the micrograph; and
(4) All tilt and rotation angles and translation elements related to a micrograph.

The specific parameters necessary for practical solutions need to be determined carefully. Limiting the distortion parameters to those just adequate for subsequent tasks can help reduce the number of control points required in the calibration procedure, and therefore, would ensure efficiency. There is, however, the possibility of encountering "critical geometry" involving physical-mathematical correlations between certain parameters, without uncoupling which no meaningful solution is possible. Ideas on critical geometry in photogrammetry are given in Ghosh (2005).

As mentioned earlier in Sec. 2.4, the best available "standards" for EM jobs are the grids made from master diffraction gratings. These being two-dimensional, a convergent configuration of two or more micrographs (of the same grid) taken by way of rotating and tilting the grid can be used to "generate" the third dimension (Fig. 19).

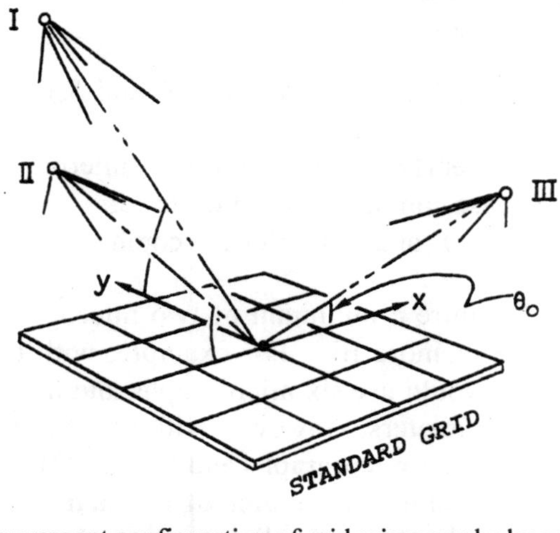

Fig. 19: Convergent configuration of grid micrographs by using one tilt (θ or ϕ) and several rotations (κ). [I, II, III are the perspective centers]

This type of configuration would give the intersection geometry at grid points as if the images are taken from different directions. Such a set-up and utilization of the process of *Self-calibration* (see Ghosh, 2005) has been found to yield the best results in such calibrations (Adiguzel, 1985; El Ghazali, 1978 and Maune, 1973). In some cases it may be advisable to use "weights" (*a priori*) for the parameters in accordance with their comparative reliabilities and as deemed fit.

The collinearity equations (Eq. 5) provide the basis for such an analytical calibration. These equations are augmented to best describe the EM geometry. The consideration of radial and spiral distortions would give the augmented version as below:

$$x_{ij} + D_1 \cdot x_{ij}^3 + D_2 \cdot x_{ij} \cdot y_{ij}^2 + S_1\left(x_{ij}^2 \cdot y_{ij} + y_{ij}^3\right)$$

$$= -C_x \frac{\left(X_j - X_{oi}\right)m_{11_i} + \left(Y_j - Y_{oi}\right)m_{12_i} + \left(Z_j - Z_{oi}\right)m_{13_i}}{\left(X_j - X_{oi}\right)m_{31_i} + \left(Y_j - Y_{oi}\right)m_{32_i} + \left(Z_j - Z_{oi}\right)m_{33_i}}$$

$$y_{ij} + D_3 \cdot y_{ij}^3 + D_4 \cdot x_{ij}^2 \cdot y_{ij} + S_2\left(x_{ij}^3 + x_{ij} \cdot y_{ij}^2\right) \qquad \text{Eq. (19)}$$

$$= -C_y \frac{\left(X_j - X_{oi}\right)m_{21_i} + \left(Y_j - Y_{oi}\right)m_{22_i} + \left(Z_j - Z_{oi}\right)m_{23_i}}{\left(X_j - X_{oi}\right)m_{31_i} + \left(Y_j - Y_{oi}\right)m_{32_i} + \left(Z_j - Z_{oi}\right)m_{33_i}}$$

where the subscript "o" refers to the perspective center; "i" refers to the photo; and "j" refers to the object point. Rest of the subscripts have been referred to in the preceding sections.

Note: The *C*'s, *D*'s and *S*'s are common to all points and photos used.

With a convergent configuration of multiple photos as indicated in Fig. 19, one finds three groups of unknown parameters:

(a) Eight parameters (C_x, C_y, D_1, D_2, D_3, D_4, S_1 and S_2) common to all points and micrographs;
(b) Two or three parameters (Z_o and the pre-tilt θ_o interpreted as ω or ϕ or $\omega + \phi$) depending on the case common to each tilt used;
(c) Three parameters (X_o, Y_o and κ) which are common to each micrograph.

A calibration would require a minimum of two micrographs (at say, 0° and 180° rotations) for each of one or more tilts. For example, with two tilt angles and three rotations (120° apart), one would use six micrographs and there would be 8 + (2 × 3) + (6 × 3) = 32 calibration parameters. However, with two tilt angles and four rotations (90° apart), there would be eight micrographs and 8 + (2 × 2) + (8 × 3) = 36 parameters. This indicates the need of using a computer of adequate capability. For many EM applications, however, consideration of radial, tangential or spiral distortions may not even be necessary. Appropriate reduction of the parameters and utilization of adequate number of "control" points in adequate number of micrographs is the key to a cost-effective calibration.

The final solution requires the formation of "observation equations", which express the relationship between the observations, the values of parameters, the errors in observations, the errors in the parameters, and the errors in the satisfaction of a particular mathematical model. Equations (19) being non-linear, are to be "linearized" by a "Taylor expansion" to obtain such observation equations in this case. The number of terms necessary in the expansion would depend on the required accuracy or the number of iterations permissible. The linearization is about the measured quantities (x_{ij} and y_{ij}) and the initial approximate values of the unknown parameters. All linearized equations, collected, may be expressed in matrix notations, concisely stated (see Ghosh, 2005):

$$\bar{V} + \bar{B}\bar{\delta} + \bar{\in} = \bar{0} \qquad \text{Eq. (20)}$$

where $\bar{V}$ is the vector of residuals; $\bar{B}$ is the matrix of the partial derivatives of the function, Eqs. (19) with respect to the parameters and the object points (grid coordinates); $\bar{\delta}$ is the vector of corrections or alterations to the parameters and object point coordinates; and $\bar{\in}$ is the discrepancy vector, $\bar{0}$ being the null matrix.

Considering a weight matrix, $\bar{W}$, of observations, by applying the principles of *least squares*, the contributions of all observations are added to what are called *normal equations*. Least squares obtains its name from minimizing the sum of the squares of the weighted residuals of all observations—that is, by minimizing $\bar{V}'\bar{W}\bar{V}$. The normal equations in compact symbolic form are

$$\bar{N}\bar{\delta} + \bar{U} = \bar{0} \qquad \text{Eq. (21)}$$

From this, one finds the solution, the correction vector:

$$\bar{\delta} = -\bar{N}^{-1}\bar{U}\,. \qquad \text{Eq. (22)}$$

In these, $\bar{N} = \bar{B}^{\mathrm{T}}\bar{W}\bar{B}$ and $\bar{U} = \bar{B}^{\mathrm{T}}\bar{W}\bar{\in}$. For interesting ideas on adjustment computations, see Hirvonen (1971) or Mikhail (1976).

3

MEASURING INSTRUMENTS

The measuring instruments are required to obtain data, either two-dimensional (*x*, *y* of image points) or three-dimensional (*X*, *Y*, *Z* of model points), with supplemental need of facilities for continuous plotting. The two principle approaches in such data acquisition are "analogical" and "analytical". The former involves optical-mechanical analog systems, while in the latter one uses computer interfaced comparators to derive and display the required information with the aid of computers. All 2-D data in EM applications are obtainable directly from micrographs. A simple *comparator* (monocular or stereoscopic; see Ghosh, 2005) would be adequate if the requirements were only to obtain photo-coordinates. The requirements being seldom limited to 2-D data only, one goes for instruments that obtain and handle 3-D data. Such instruments are all capable of reading 2-D data as well. Furthermore, the necessity of single image analogical instruments like rectifiers or sketch masters have never surfaced in EM applications.

The analog and analytical instruments, each has three basic components:

(i) The viewing system;
(ii) The measuring system (including orientation mechanisms); and
(iii) The readout and recording system.

The acquired data, after possible corrections and manipulations, may be directly displayed graphically, with a mechanical analog plotter, or handled digitally, leading to computer mapping.

3.1 ANALOGICAL INSTRUMENTS

Because the EM imaging systems have parallel or very nearly parallel projection and most analogical photogrammetric plotters have perspective projections (optical or mechanical), they are greatly incompatible. The primary effect of using such analog plotters would be significant scale-affinity in the stereo model, affecting the *Z* coordinate as compared with the *X* and *Y*. One way of alleviating this problem is to determine the appropriate scale factor for the heights (*Z*) of points and utilize developed (ad hoc) nomograms for continuous mapping of features as was done by Oshima *et al*. (1970) working at a Wild A7 Autograph. A second approach would be to modify such a stereoplotter for EM applications. A successful modification of a mechanical projection type instrument (Wild B9 Aviograph) was reported by Wood (1972). This would be impossible with a direct optical projection type analog instrument. The third possibility of using approximate (with no rigorous perspective projection) instruments having the capability of correcting the deformed model (such as Zeiss Stereotope) has been demonstrated by Ghosh (1971). Such a system interfaced with a computer and a plotter gives virtually an analytical plotter and has been found to be well suited to EM applications. The only other possibility is using specialized instruments made specifically for EM applications such as the EMPD (see details below). These specialized ones, however, have their inherent optical-mechanical limitations. A precision analytical plotter, or at least an analog-analytical hybrid, therefore, seems to be the ultimate answer to all problems of EM applications. Representative ideas on these are given below:

A. Instruments Specifically Designed for EM Applications

The EMPD (Electron Micrograph Plotting Device) Model 2 (Fig. 20) is a good example. It was developed specifically for the SEM by Cardiographic Engineering Ltd., Salisbury,

Fig. 20: Electron Micrograph Plotting Device (EMPD), Model 2
[*Courtesy* of The Commonwealth Scientific Corporation, Alexandria, Virginia, USA]

UK and marketed in the USA by Commonwealth Scientific Corporation of Alexandria, Virginia. Its use with success was reported by Boyde and Ross (1975). The design is around two basic assumptions, that a simple optical-mechanical solution is adequately efficient, and that distortion-free photographs are available. Based on parallel projection in the EM and the same magnification in each micrograph of the stereo pair, the instrument is designed such that the scale variation in one photo, assumed to be due to tilt only, is rectified by using a cylindrical lens to cause an anamorphic change. The intersection at a point yields the scale-corrected *parallax* ($\Delta\ p$), such that if one chooses the left side photo as the datum (reference plane) for measuring heights, the right side photo is foreshortened by $\cos\theta$ so that $\Delta\ p = x' \cdot \cos\theta - x''$ (analogy from Eq. 18). This gives the following expression for height:

$$Z' = \Delta\ p/(\sin\theta \cdot M) \qquad \text{Eq. (20)}$$

The freehand motion in plotting is coordinated with the movement in the model of the floating mark, similar to a standard photogrammetric plotter. Manual plotting in different scales is possible with the help of a pantograph on the plotting table. The instrument can use micrographs of formats upto 100 × 125 mm. It has the capability of plotting profiles and contours anywhere in the model and in any direction. The optimum parallactic angle to be used is 10° as suggested by the manufacturer.

B. Standard Analog Photogrammetric Stereoplotters

Any precision instrument (like Wild A10 or Zeiss C8) designed for plotting and extraction of 3-D data from overlapping photographs can be used. One such, with auxiliary devices interfaced with digitizers, etc. would be adaptable to various demands and applications with EM. An example, the Wild (current name, Leica) B9 Aviograph, (see Fig. 21) was used with success as reported by Wood (1972). In this, the two cameras (projectors) can be oriented at will. It can be also used as a mono- or stereo-comparator.

The Wild B9 can accommodate all sizes and forms of negatives or diapositives on films or plates upto 11.5 × 11.5 cm. Focal lengths of 44 ± 3 mm can be set with an accuracy of 0.01 mm. The manufacturer can provide picture carriers to compensate for known radial lens distortions. Spatial control of the measuring mark is done by two handwheels (for X and Y) and a footdisk (for Z) movements. The observation optics has a 7.5 × magnification, and the binocular is adjustable to individual (eye-base) requirements. For relative orientation of the model, each picture can be inclined through ± 5g in φ and ω tilts, and through ± 15g in κ rotation. For absolute orientation, the model base can be varied between 38 and 184 mm (to effectuate scale change) and inclined about its center point by ± 5g. With auxiliary equipment, additional profiling, printing of coordinates, etc. are possible with high precision. With interfaced digitigers, this instrument would develop into an automated cartographic capability with the DTM (Digital Terrain Modelling) Technique.

Fig. 21: Wild B9 Aviograph (with additional table and pantograph) [*Courtesy* of Leica Geosystems, Switzerland]

Since the EM geometry involves parallel projection, it is highly desirable that one sets the largest possible focal length (principal distance) at such an instrument. The perspectivity of the instrument effects a model-deformation, which would give falsified heights. The height of a point *P* in the stereo-model with regard to a reference point *B* (Fig. 18) is given by (refer to any standard book on photogrammetry):

$$Z' = \frac{b \cdot f(P_1 - P_2)}{P_1 \cdot P_2 \cdot M} = \frac{B \cdot f \cdot \Delta P}{P_1 \cdot P_2 \cdot M} \qquad \text{Eq. (21)}$$

where b is the base setting;

f is the focal length;

P_1, P_2 are the parallaxes at points *B* and *P*, respectively, Δp being the parallax difference;

and M is the magnification.

The parallaxes of the two points can be regarded as known or fixed quantities. The base and the focal length (b and f), being quantities of no physical significance in parallel geometry, can be chosen arbitrarily in view of the parallactic angle (γ). One can infer that with proper choice of b and f, the height difference can be reduced to zero. This is true if one is interested only in the height difference between two points in the object. However, in 3-D continuous mapping, one is faced with the inevitable model deformation (due to affinity in height), which can be kept within certain bounds by proper use of Eq. (21). If the base and focal length values are chosen rather arbitrarily, in some cases the height exaggeration can be unmanageable. In order to avoid this, it would be advisable to establish the relationship between the Z' values of parallel and perspective projections for the specific values of b and f used (by considering Eqs. 17 and 21 as was done by Oshima *et al.*, 1970).

In this respect, in can be recommended that, for the use of such a stereoplotter, a table be prepared showing the numerical value relationships among the parameters P_1, P_2, Z'_{para} and Z'_{pers} for the particular combination of b and f. Next, by using them, a correction graph be constructed by way of plotting Z'_{para} against Z'_{pers}. Such graphs can be used in practice or the relationship can be programmed for all mapping jobs with the same base and focal length settings at the same instrument (El Ghazali, 1978).

It is possible that slight corrections would be necessary in the X' and Y' coordinates. In most cases, however, with no lateral (ω) tilt, Y' does not change and if the parallactic angle is sufficiently small, a unit value may be considered for the cosine function in X'. Thus, both X' and Y' can be taken directly from one of the two photos (with due consideration of the magnification). This will be more so if the left side photo is vertical (i.e., zero tilt) and the right side one has the desired tilt to create stereo view. This would, of course, tilt the datum for the Z' values. For most applications, this is no cause of concern as there is no unique or natural datum in EM applications. See Boyde and Ross (1975) for an interesting discussion on the choice of datums in SEM applications.

C. Hybrid (or Converted) Analog Plotters

Conversion of analog plotters into analytical has been successfully accomplished during the 1980s and later. Examples are provided in Qasco (QA) conversion of Wild B8 and Kern PG2 interfaced to a "mini-computer" (Fig. 22) and E. Coyote's conversion of several Topocarts (Anacarts) by using computers (like DEC PDP-11). Such low cost PC based hybrid plotters are becoming popular world-wide and would be very suitable for EM applications. In the QA conversion, for example, the existing tracing stand and space rod systems are removed and replaced with microprocessor controlled servo-motor drive systems incorporating a digital encoder feedback loop. The system has a least count of 2 μ m at photo and, depending on the optics and floating mark size, would give an accuracy of the order of 5μ m. The focal length range being unlimited, it has no problem with parallel projection geometry. The maximum scanning area is the same as in the original analog plotter (24cm × 24cm) and the scanning speed (variable) is from zero to 25 mm per second. Operator controls are versatile, with handwheels, mouse, joystick, foot disc or *X-Y* digitizer or combination.

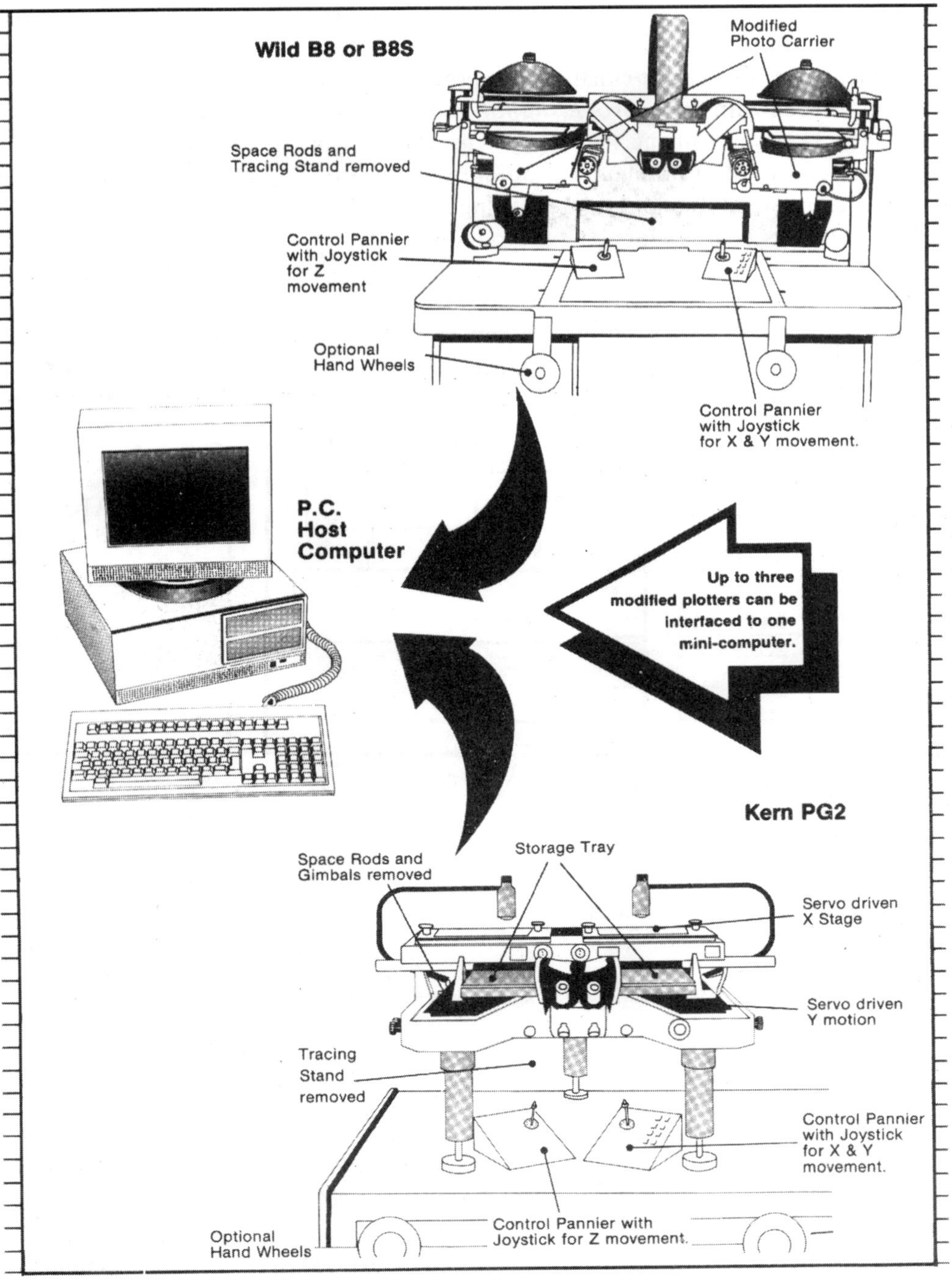

Fig. 22: Qasco (QA) conversion system diagram (with Wild B8 and Kern PG2)
[*Courtesy*: Qasco Analytical Systems, Australia]

3.2 ANALYTICAL INSTRUMENTS

An analytical plotting system consists of a precision *stereocomparator* and a *coordinatograph* interfaced with an *electronic digital computer* plus other auxiliary devices as may be necessary. These instruments (see the system schematic in Fig. 23) are capable of solving wide variety of photogrammetric problems. Operations like relative and absolute orientations, continuous mapping and profiling are possible in such instruments, apart from these being used as comparators (stereo- ro mono-).

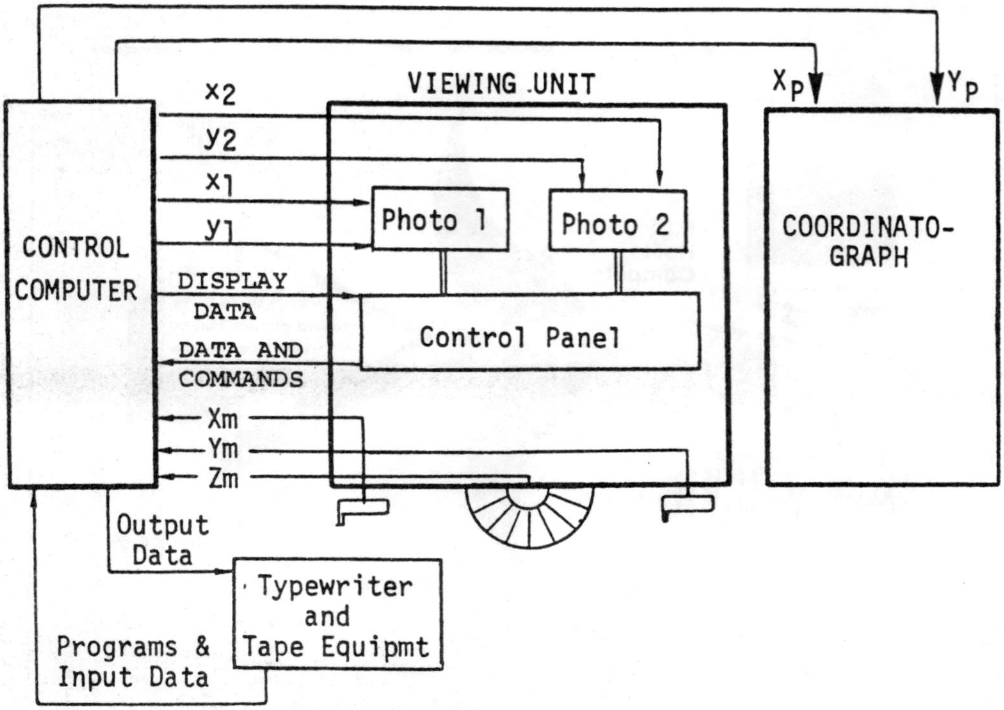

Fig. 23: System Diagram of an Analytical Plotter

A. Standard Analytical Plotters

The most sophisticated systems in this group are categorized as those of the "first order", like AP/C-4 (by OMI SpA of Rome, Italy), Zeiss C-130 Planicomp (by Zeiss, Oberkochen, Germany), US-2 (by Helava Associates, USA) or Wild BC3 (by Leica Geosystem, Heerbrugg, Switzerland; see Fig. 24). Their sophistications comprise features like: (a) all plotting applications; (b) measuring digital terrain models (DTM); (c) photo-triangulation for strips and blocks; and (d) other data processing for various derived information. Further extension of such sophistications in recent years have produced another version known as

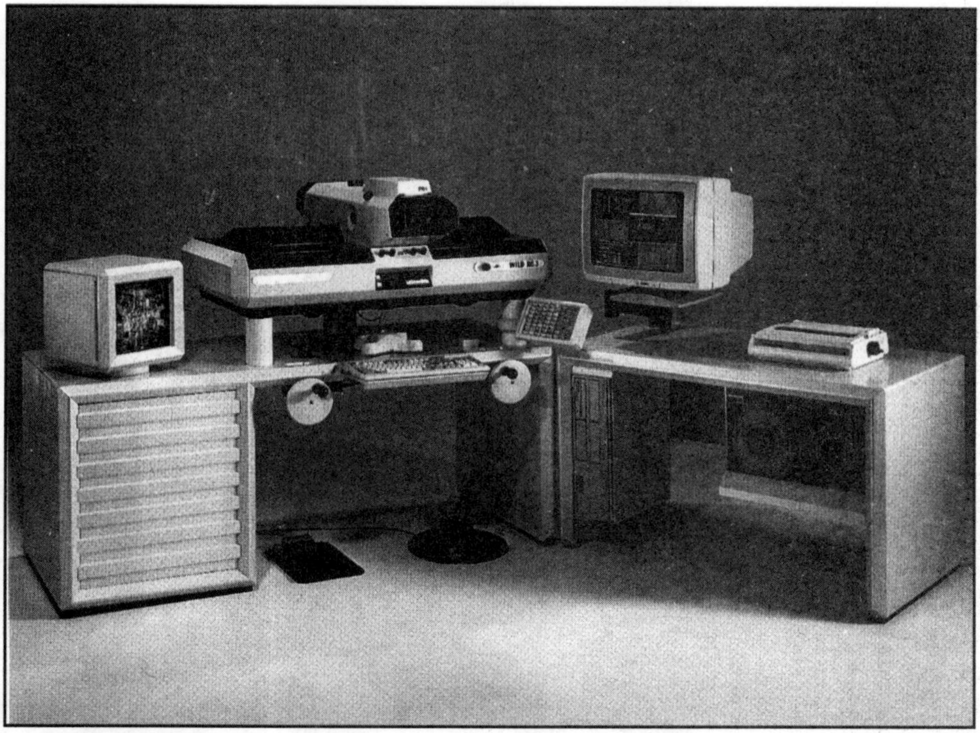

Fig. 24: Wild BC3 Analytical Plotting System
[*Courtesy* of Leica Geosystem, Switzerland]

Digital Photogrammetric Workstations (DPW) or Softcopy Workstations (SW) such as one developed by LH Systems, Inc., San Diego, CA or one by Intergraph Corporation, Huntsville, AL. Such sophistications are not at all necessary in EM applications. In this regard, analytical plotters of the "second order", like Zeiss P2 Planicomp (by Zeiss, Oberkochen Germany) or Digicart 40 (by Galileo Siscam, Florence, Italy; see Fig. 25) would be ideally suitable. With appropriate modifications, the following tasks (and more) can be routinely performed at such instruments:

1. Interior orientation, with due consideration of appropriate projection, distortions, and dynamic corrections (if any), in each photograph.
2. Relative orientation of the stereo pair.
3. Absolute orientation of the stereo model.
4. Cartographic presentation of the data at the plotting table (coordinatograph), including drawing of profiles in any direction.
5. Recording and storing of digital information.
6. Obtaining various derived information such as area, volume and perimeter.

Fig. 25: Digicart 40 Analytical Plotter
[*Courtesy* of Galileo Siscam, Florence, Italy]

B. Simple Analytical Plotters

Several analytical plotters with limited scopes (e.g., limitations in computer memory or in the design) have recently appeared on the market or have been developed by individuals on ad-hoc basis. The APPS-IV (marketed by Autometric, Inc., Arlington, Virginia, USA) and Stereocord G2 (by Zeiss, Oberkochen, Germany) are two outstanding example. These are amazingly suited for most EM applications. Because of the limitations, these are comparatively low priced and yet reasonably precise. The Stereocord G2 (see Fig. 26) is, in reality, a modified Stereotope (by Zeiss), its analog mechanical rectiputers having been replaced by an electronic desk calculator.

Fig. 26: Stereocord G2 System (with Direc-1, desk-top computer HP 9830, Printer 9866A and Planimetric Plotter HP 9872A)
[*Courtesy* of Zeiss, Oberkochen, Germany]

This instrument has linear encoders on x, y, Δx and Δy (i.e., the photo coordinates and the two parallaxes) motions. These measured values are digitized directly with the aid of a commercial pulse generator in conjunction with a counter (Direc-1) and an interface system. All numerical data for each stereo pair are stored together with the computer program required for data reduction. The Stereotope part acts as a stereo viewing comparator. All data reductions are performed automatically at the computer and the 3-D coordinates can be printed and displayed. When the plotter is connected, it becomes a complete analytical plotter. One such instrument system has been used with success in plotting more than two thousand SEM and TEM stereo models (see Ghosh *et al.*, 1978). This system was found to be about ten times as fast in 3-D data acquisition, although about half as much poorer in the obtained accuracy, than a precision analog stereo plotter like Wild A10.

C. Comparators

Mono- or stereo-comparators (like Zeiss PSK or Wild STK), adequately fitted with accessories for registration of digital data to be used off-line at a computer and associated plotting facilities would be the other possibility. With appropriate mathematical models, this approach would give results similar to those with precision analytical plotters. Being off-line, however, this approach would be less economical.

4

ACCURACY AND RELIABILITY

Technological, economic and circumstantial factors interact in a complicated fashion in designing and performing a project. Certain fundamental technical considerations are necessary in designing and evaluating any project relating to the extraction of quantitative data. In EM applications, comparative (or relative) results from different specimens are often more important than absolute accuracy. The variability between specimens sometimes far outweighs the importance of some accuracies in individual results. This leads to a general reluctance to consider accuracy related aspects. Nonetheless, they are important considerations and are, therefore, briefly presented here. Ideas are drawn from the domains of topographical and engineering mapping, which are standard jobs in menstrual data handling. Standard accuracy specifications and evaluation (testing) criteria are usually established for standard jobs. For elaborated ideas on these, the reader may benefit from various publications on Adjustment Computations (e.g. Wolf and Ghilani, 1997).

A. Accuracy

In general discussion on *Accuracy*, we must specify both, a range of values within which we expect the output to lie and the probability that will occur. There are two other terms used in this regard, viz., *Bias* and *Precision*. *Bias* is the difference between the input and the mean (accepted as correct) value of the output. *Precision* is the degree of mutual agreement among individual measurements (made under prescribed like conditions), thus it describes the repeatibility issue also. In this light, *Accuracy* is defined as the degree of agreement of individual measurement with an accepted reference value. There is general agreement on the desirability of using the term, *Standard Deviation*, *Standard Error*, or *Root-mean-square-error*, to express accuracy directly or indirectly. Depending on the circumstances, however, the concept must be considered in terms of the following five possibilities in EM applications, for which the following notations are used ("*i*" refers to individual observations):

$t_1, t_2, \ldots, t_n$	observations; n is the number of measurements;
$x, y, z, \ldots$	unknowns; u is the number of unknowns;
$v_i = x - t_i$	corrections in direct observations;

$v_i = f(x, y, z, \ldots) - t_i$	corrections where unknowns are related by functions (linear or nonlinear);
$m_x, m_y, m_z, \ldots$	standard errors (deviations) of unknowns;
m and m_0	standard errors of one observation and one observation of unit weight, respectively;
p	weight of one measurement.

Possibility 1. Adjustment of Direct Observations of Identical Accuracy

In this case, $p = 1$; $x = (1/n)(t_1 + t_2 + \ldots + t_n) = [t]/n$ and $[v] = 0$ are assumed. These give, from basic statistical principles,

$$m = \pm\sqrt{[vv]/(n-1)} \quad \text{and} \quad m_x = \pm m/\sqrt{n} \qquad \text{Eq. (22)}$$

Possibility 2. Adjustment of Direct Observations of Different Accuracy

In this case, one assumes

$$p = m_0^2/m_i^2; \quad [vp] = 0$$

and

$$x = \frac{t_1 p_1 + t_2 p_2 + \ldots + t_n p_n}{p_1 + p_2 + \ldots + p_n} = [tp]/[p]$$

These give

$$m_0 = \pm\sqrt{[vvp]/(n-1)} \qquad m_i = \pm m_0/\sqrt{p_i} \qquad m_x = \pm m_0/\sqrt{[p]} \qquad \text{Eq. (23)}$$

Possibility 3. Adjustment of Linear Functions (in View of the Law of Error Propagation)

One assumes $x = a_1 t_1 + a_2 t_2 + \ldots + a_n t_n$, the a's being certain constants.
These gives

$$m_x^2 = a_1^2 m_1^2 + a_2^2 m_2^2 + \ldots + a_n^2 m_n^2 \qquad \text{Eq. (24)}$$

Possibility 4. Adjustment of Nonlinear Functions (in View of the Law of Error Propagation)

One assumes $x = f(t_1, t_2, \ldots, t_n)$. This gives

$$m_x^2 = \left(\frac{\partial f}{\partial t_1}\right)^2 m_1^2 + \left(\frac{\partial f}{\partial t_2}\right)^2 m_2^2 + \ldots + \left(\frac{\partial f}{\partial t_n}\right)^2 m_n^2 \qquad \text{Eq. (25)}$$

Possibility 5. Adjustment of Intermediate Observations (in View of the Law of Error Propagation)

The original error equation can be set up in the form

$$T_i + v_i = f_i(x, y, z, \ldots) \quad \text{with weights, } p_i = m_0^2/m_i^2$$

Next, approximate values of the unknowns can be introduced: $x = x_0 + dx$; $y = y_0 + dy$; etc. Consequently, the approximate function values can be computed:

$$f_i(x_0, y_0, z_0, \ldots)$$

Then the error equations can be written in absolute terms:

$$-t_i = -\{T_i - f_i(x_0, y_0, z_0, \ldots)\}$$

The transformed error equations can then be set in the following form:

$$v_i \cdot \sqrt{p_i} = a_i\sqrt{p_i}\, dx + b_i\sqrt{p_i}\, dy + c_i\sqrt{p_i}\, dz + \ldots - t_i\sqrt{p_i} \qquad \text{Eq. (26)}$$

where the coefficients are $a_i = (\partial\ f_i/\partial\ x)_0$; $b_i = (\partial\ f_i/\partial\ y)_0$; $c_i = (\partial\ f_i/\partial\ z)_0$, etc. The changes (corrections), *dx*, *dy*, *dz* etc., are computed from the "normal equations" set after Eq. (26). The standard deviations (errors) are given by

$$m_0 = \pm\sqrt{[vvp]/(n-u)} \quad \text{and} \quad m_x = m_0\sqrt{q_{xx}}$$

$$m_y = m_0\sqrt{q_{yy}} \quad m_z = m_0\sqrt{q_{zz}} \quad \text{etc.} \qquad \text{Eq. (27)}$$

where the q's are the weight reciprocals (from variances).

Discussion of the law of error propagation, the principles of least squares, the formation and solution of normal equations, etc., can be found in standard books [e.g., Wolf and Ghilani, 1997].

B. Planimetric Plotting

There is widespread agreement for separating planimetry (by combining X and Y coordinates) from height (Z coordinate). There is, however, a continuing debate whether the planimetric accuracy should be expressed in terms of univariate or bivariate (consisting of two variables, X and Y) data.

In keeping with common practice in mapping, the *linear standard error* for the X coordinate is

$$\sigma_X = \left\{ \frac{\sum_{i=1}^{n} (\Delta X_i)^2}{n-1} \right\}^{1/2} \qquad \text{Eq. (28)}$$

where $\Delta\ X_i$ are the coordinate discrepancies between accepted (as true or absolutely correct—for example, from calibrated grid observation data) and mapped positions at each test point; and n is the total number of points used for checking. Similarly, linear standard errors for the Y and Z coordinates are obtainable. It is also debatable whether n or $n - 1$ should be used in the denominator of Eq. (28) For large values of n, however, the difference is not significant.

Next, the *planimetric standard error* (σ_P^2) is expressed by

$$\sigma_P^2 = \sigma_X^2 + \sigma_Y^2 \qquad \text{Eq. (29)}$$

Some prefer to use the *two-dimensional (circular) standard error* in such cases. This is the error in a quantity difined by two random variables, with the basic assumption that planimetric errors are expected (in continuous mapping) with equal probability in any direction and with equal magnitude, which is a very popular concept with many practitioners. The circular standard error (σ_c) is related to σ_p as shown in Eq. (30).

$$\sigma_c = \sigma_p / \sqrt{2} \qquad \text{Eq. (30)}$$

For interesting ideas in this regard, see Thompson and Rosenfield (1971).

C. Contour Plotting

The permissible minimum contour interval should relate to the obtainable accuracy in height. A relationship widely used in the world of mapping is:

$$\sigma_Z = 0.3\, C \qquad \text{Eq. (31)}$$

where σ_Z is the standard error in Z obtainable from the working system, and C is the permissible minimum contour interval.

Note: The factor 0.3 in Eq. (31) is based on the fundamental assumption that, with normal distribution of errors, spot heights as derived by interpolation from contour lines are expected to be in error by more than σ_Z at not more than 10 per cent of the points. Further discussion on contouring is given by Richardus (1973). A very logical question still remains concerning whether correct statistical theory is applied in using this factor. The practitioner, however, may have other strong reasons for selecting a specific contour interval.

D. Stability and Repeatability

Stability and repeatability are two interrelated words often used to express the same meaning. Certain ideas have already been presented in Chap. 2. Below are given additional, accuracy related, ideas on these topics. Leipholz (1970) postulates that first an "unperturbed state" of the system be specified. Next, a perturbation is applied by repeating the same study but changing only the specific "element" or "parameter" being evaluated. Certain statistical tests would characterize the state of the system. Alternatively, without changing the elements or parameters, the system can be evaluated in terms of the output under repeated performance.

Stability tests (several) on parameters or elements in EM systems (Maune, 1973; Nagaraja, 1974; and El Ghazali, 1978) indicate that, with the usual consideration that 90 per cent of checked items should pass a test, even at the significance level of 0.02, the mensuration related elements remain statistically stable.

Repeatability tests on *X*, *Y* and *Z* coordinates generally indicate results consistent with those obtained after the parameters have been tested (El Ghazali, 1978; and Adiguzel, 1985). These results appear to be even better, usually being statistically stable at the significance level of 0.01 for all three coordinates.

5

APPLICATION TO SINGLE SPECIMENS

The idea of using TEM for stereoscopic measurements of specimen features occurred at an early stage in its development (Helmcke, 1954). Later, the TEM was believed to provide a parallel projection, at least when used at high magnification (Helmcke, 1965). Simple geometric relationships (as in Eqs. 17) were found to yield satisfactory results particularly if the two micrographs were obtained by tilting the sample symmetrically. Analogical-graphical mapping for planimetry (sometimes even by tracing the photos directly) and a separate solution for spot heights determined by means of the parallax equation would suffice in most cases.

Initially, the photogrammetric work with SEM did not consider solutions to be much different from those with TEM. It was, however, recognized (Wells, 1960) that the projection is evidently perspective, at least at very low magnifications, where the micrographs are the sharpest, and the errors arise from the use of false assumptions in the projection. Since the TEM and the SEM offer different types of information, it will be appropriate to study their applications to single specimens separately. On the other hand, in certain specimens, a combination of information from both would yield data the equal of which is impossible with any other system. This possibility will be discussed in the third section.

5.1 SEM APPLICATION

In the SEM, the focussed beam impinges upon the specimen surface. The electron radiation is used for two synchronous scanning beams, one sweeping over the surface and a corresponding second one over a cathode ray tube (CRT), which is recorded with a camera. The micrograph thus obtained shows the specimen surface features. Somewhat low resolution (~5 nm) and the absence of subsurface feature are the two primary characteristics of SEM micrography.

Most SEM users find it convenient to use cameras capable of recording the specimen almost instantaneously (for example, with a Polaroid Land Camera). These cameras are designed for recording the CRT displays at 1:1 or lower scales. The total magnification, however, refers to the image recorded by the camera. Therefore, the 3-D mapping cannot always be referred to the scale of the visual (CRT) display. Moreover, with the polaroid type cameras, although paper prints may be convenient, for the high-precision mapping one should insist on glass plates or film. There are two reasons behind this recommendation. First, paper as the emulsion carrier is seldom adequately stable dimensionally. Second, most precision photogrammetric measuring instruments are designed for transparencies.

As has been indicated earlier, resolution in the SEM system plays a major role in 3-D mapping. A 5 nm resolution with magnification of 10,000x would correspond to 0.05 mm on the micrograph. The measuring instrument having a standard error of less than 0.05 mm on the photo should then be considered as adequate. Higher magnification would tend to result in unsharp images. Lower magnification would give sharp photos but would demand more precision instrument, a more stable photo base, etc. Therefore, depending on the SEM and the measuring instrument, an optimum magnification (with regard to the desired accuracy) needs to be determined.

The following working steps are recommended for meaningful 3-D mapping with SEM micrography:

1. Determination of the optimum magnification and parallactic angle.
2. Consideration of the distortions in the micrographs, the method and extent of their elimination (or decrease).
3. Consideration of the mathematical model(s) with regard to the specific instrument, the magnification, and the degree of refinement desired.
4. Obtaining the necessary pair of micrographs at the SEM (with due consideration of the emulsion carrier).
5. Performing the orientations (Interior, Relative and Absolute) at the stereo-plotting instrument, or observations at a comparator and subsequent data reduction at a computer.
6. Obtaining 3-D data and/or graphical-cartographic presentation of data.
7. Obtaining other necessary derived data.

Note: In using an instrument with limitations in projections with micrographs (like an analog plotter), there is bound to be an appreciable scale-affinity in Z (height) as compared with X and Y (planimetry). Accordingly, the height values need to be appropriately corrected for in absolute orientation as well as in contour mapping.

An example of a contour map plotted at a Wild A7 stereo plotter with a pair of SEM micrographs in illustrated in Fig. 27.

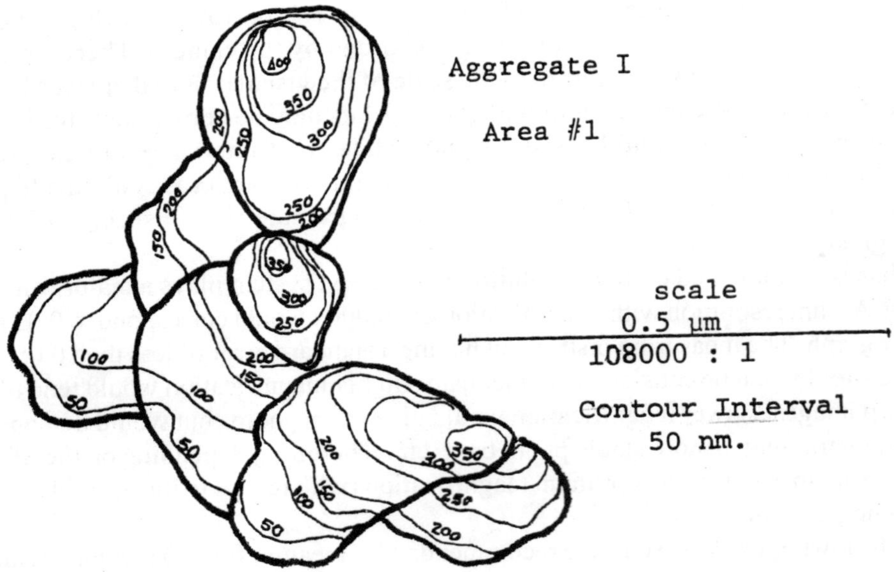

Fig. 27: Contour map of a carbon black aggregate with a pair of SEM micrographs at a Wild A7 Autograph.

Scale of micrographs: 12k x; scale of stereo model: 27k x; scale of contour map: 108k x [*Note*: The scale change in this Fig. is due to reproduction]. Stereo (parallactic angle: 15°; Pretilt (left photo): – 7°30′; Contour intervals: 50 nm. The thick lines indicate grouping of particles for specific investigations in research.

5.2 TEM APPLICATION

In the TEM, for obtaining a micrograph, the electron beam is transmitted through the specimen (an ultrathin section), defocussed below the specimen, and projected on the photographic emulsion surface. High resolution and loss of surface features are two typical characteristics of the TEM micrographs. If 3-D data of the sample are to be obtained, it must be no larger than the thickness of the specimen (slice). The magnified image can also be displayed on a fluorescent screen (CRT).

Basic photogrammetric considerations of optimum magnification, optimum stereo angle and the working steps for mapping are the same as for the SEM applications. However, because it has a much better resolution as compared with the SEM, stereo plotting with its micrographs offers problems much less than with the SEM. Figure 28 illustrates the cartographic representation of the same carbon black aggregate as shown in Fig. 27 performed at the same Wild A7 Autograph with a pair of TEM micrographs. Depending on the specimen, contour mapping in the true sense of the term may be often impossible, as in this case where the particles in the TEM stereo appear as disks with no features on their surfaces. However, height determination of specific identifiable points or levels poses no problem.

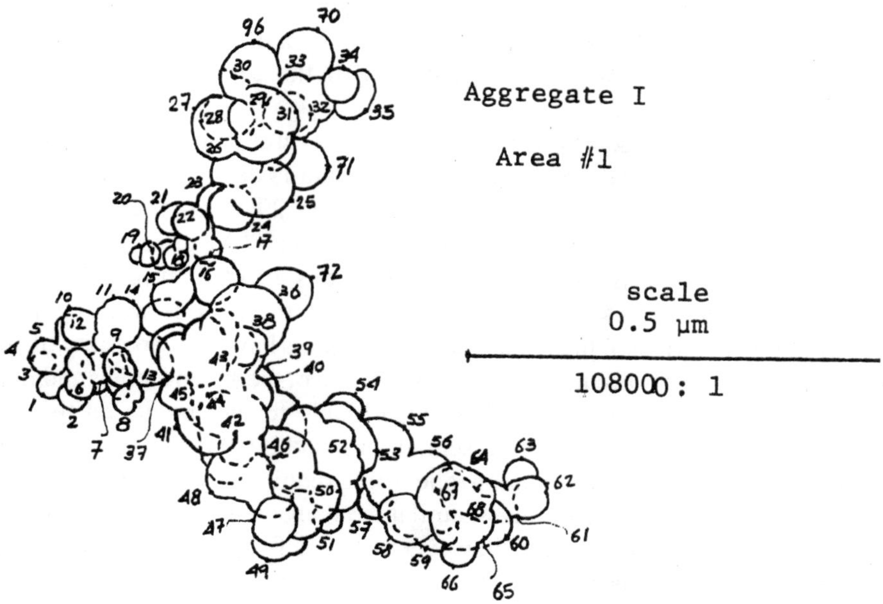

Fig. 28: Cartographic representation of the carbon black aggregate (same as in Fig. 27) with TEM micrograph pair at the Wild A7 Autograph.

Scale of micrographs: 18k x; scale of stereo model: 36k x; scale of plotting: 108k x [*Note*: The scale change in this Fig. is due to reproduction]. Stereo angle: 15°; Pretilt (left photo): – 7°30′. The numbers are for reference to a list of spot heights (not reproduced here). The broken lines indicate segments of perimeter lines below other particles and yet are visible in stereo, owing to the nature of the TEM.

Since TEM micrographs provide no background features, orientations of the stereo model by using points outside the specimen may puzzle an uninitiated photogrammetrist. This difficulty can be circumvented by using points at or near the perimeter of the specimen and by applying analytical (numerical) orientation procedures rather than empirical ones. In the same context, one has to select a specific point as the height datum (zero height). In Fig. 28, the point number 1 (at the edge of the lowest particle of aggregate) refers to the height datum.

The problem of appreciable scale affinity in the *Z* coordinate occurring in the use of perspective projection type stereo plotter (as in SEM) would exist in this also. Appropriate corrections are, therefore, necessary. Without due consideration of such affinity in the stereo model, the necessary rotations of the model may not be adequately correct and the final 3-D data would reveal unusual deformations, adjustment of which may simply not be easy.

5.3 COMBINING DATA FROM BOTH, SEM AND TEM

In certain cases, an amalgamation of data from SEM and TEM for the same specimen would result in the best possible representation of the dimensions and related information.

This simply involves superposing the data from one stereo model on to the other. It can be done conveniently either at the stage of the stereo model or at the stage of graphical plotting. This also assumes the same absolute orientation of the two models at the stage of amalgamation.

The two imaging systems being different, this requires careful planning for the extraction of 3-D data. When an empirical-graphical display of data is desired, it is best done at the stage of final plotting. One such example is presented in Fig. 29, where the data from Figs. 27 and 28 are combined. The same combination of SEM and TEM plots with technique of shading (an artistic rendition) is presented in Fig. 30. In this effort, after initial trials, in view of the desired scientific studies, scale considerations for the various working steps were established as in Fig. 31. Most of the precision stereo plotters have the capability of adjusting to such flexibility. Adequately accurate absolute orientation of the stereo models and/or plotting in such applications is imperative. Distinctly identifiable feature points on the specimen common to the two (SEM and TEM) are essential for matching the data. Assuming that the individual models are free from deformation, one would require only three widely distributed points on the specimen. In principle, this is similar to an absolute orientation with regard to three control points. With more points, however, a better fit is possible, its quality being dependent on the number, distribution and accuracy of locating such points. In case digital terrain modelling (DTM) capability is in the working system, such superposition is better done at the stereo model. Scaling, rotating and translating one model with respect to the other are than operations that are better performed with a high speed computer. An interfaced plotter would then complete the job of displaying the desired information.

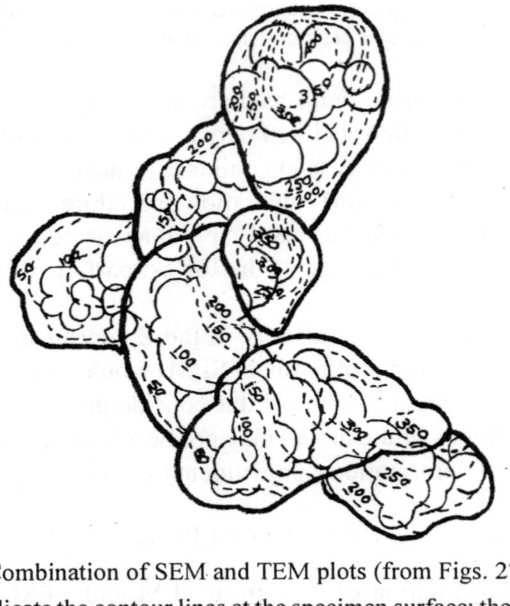

Fig. 29: Combination of SEM and TEM plots (from Figs. 27 and 28).

[*Note*: The broken lines indicate the contour lines at the specimen surface; the rest are as described in the respective figures]

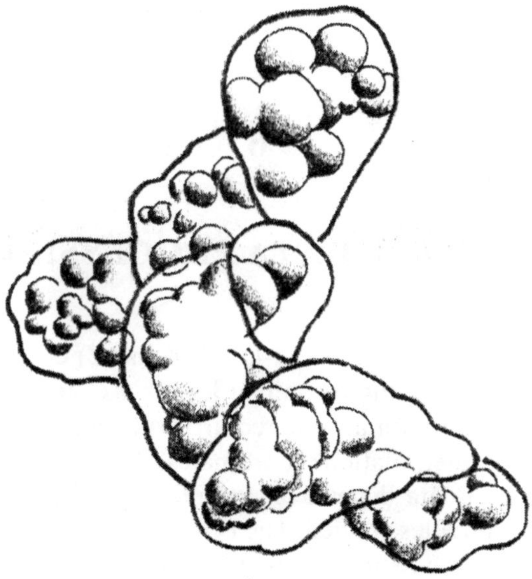

Fig. 30: Combination of SEM and TEM plots (same as in Fig. 29) with shading Technique

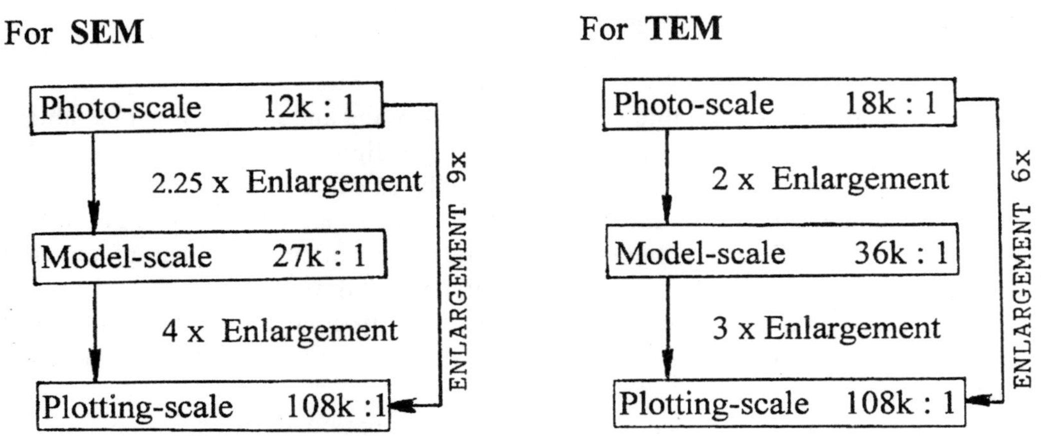

Fig. 31: Scale considerations at various working steps with SEM and TEM

6

DIGITAL DATA PROCESSING

The surface of an object in terms of the 3-D (*X*, *Y* and *Z*) coordinates of points can be described numerically by using the technique known as the *Digital Terrain Modelling* (DTM). This is usually done with an ordered array of numbers, which can be based on measured and derived (by interpolation) coordinates of discrete points on the surface. The derived (via interpolation) data depend on the measured (actual observations) and the corresponding interpolation rules applied. The DTM concept has been widely used in many engineering applications since its inception at the Massachusetts Institute of Technology (Miller and Laflamme, 1958). The *Z* coordinate being analogous to elevation (height), it is also widely known as *Digital Elevation Modelling* (DEM) in some publications related to mapping.

Any work involving DTM has two essential phases:

(1) *Data sampling* (acquisition); and
(2) *Data interpolation* (processing).

Both require a considerable amount of data storage and management capabilities, performed essentially with a computer. Once the digital data have been generated, further numerical, statistical and graphical analyses can be made through the use of high speed computers. For example, computation of surface area, volume of the object, slopes, curvatures, distances, angles, perimeters, centers of gravity and graphical outputs like contour cross-sections (profiles) and perspective drawings can be easily derived from the DTM data bank. Such tasks are conveniently done at an analytical plotter or a Digital Photogrammetric Workstation (DPW). Figure 32 describes the schematic of such a system.

The efficiency of DTM depends on the following basic factors, viz.,

(*a*) The type of surface;
(*b*) The density and distribution of terrain points used;
(*c*) The desired and obtainable accuracy; and
(*d*) The type of interpolation method used.

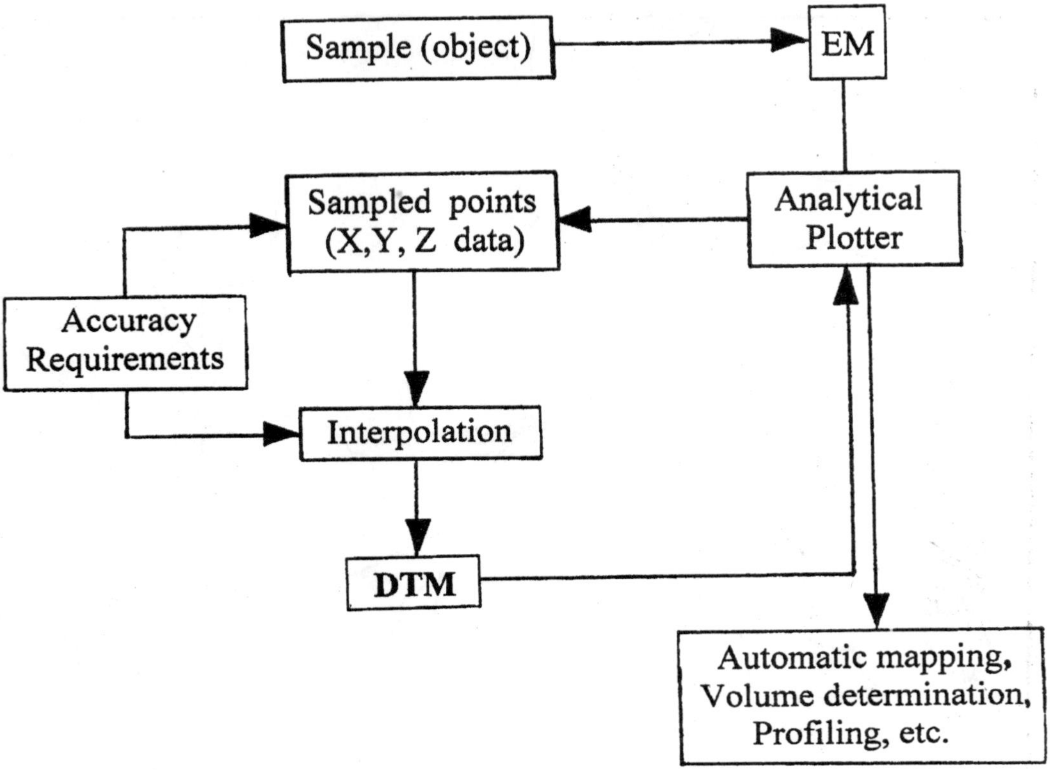

Fig. 32: Role of DTM with regard to an analytical plotter

There is an understanding in the scientific community that DTM data have much more value than just mapping applications. One major benefit is the ability to store the data for use on a real-time basis or for statistical or other analyses.

Modern technology is advanced enough that computer mapping of specimens and corresponding perspective diagramming are routinely possible. Examples of such applications are given in Figs. 33, 34, and 35. These are self-explanatory. For interesting ideas on interpolation methods in DTM, see Schut (1977) and Schenk (1999).

The DTM approach also offers many possibilities for qualitative and quantitative studies of dynamic objects by way of *differential mapping,* that is, the mapping of an object surface relative to itself in view of changes in time and space or relative to another object with which it may bear some physical and dimensional relationships. An example is given in Fig. 36.

One of the major goals of the DTM approach has been, and will continue to be, the automatic generation of DTMs. Considerable progress has been made during the last decade, but many hardware and software problems persist, preventing "robust" solutions yet.

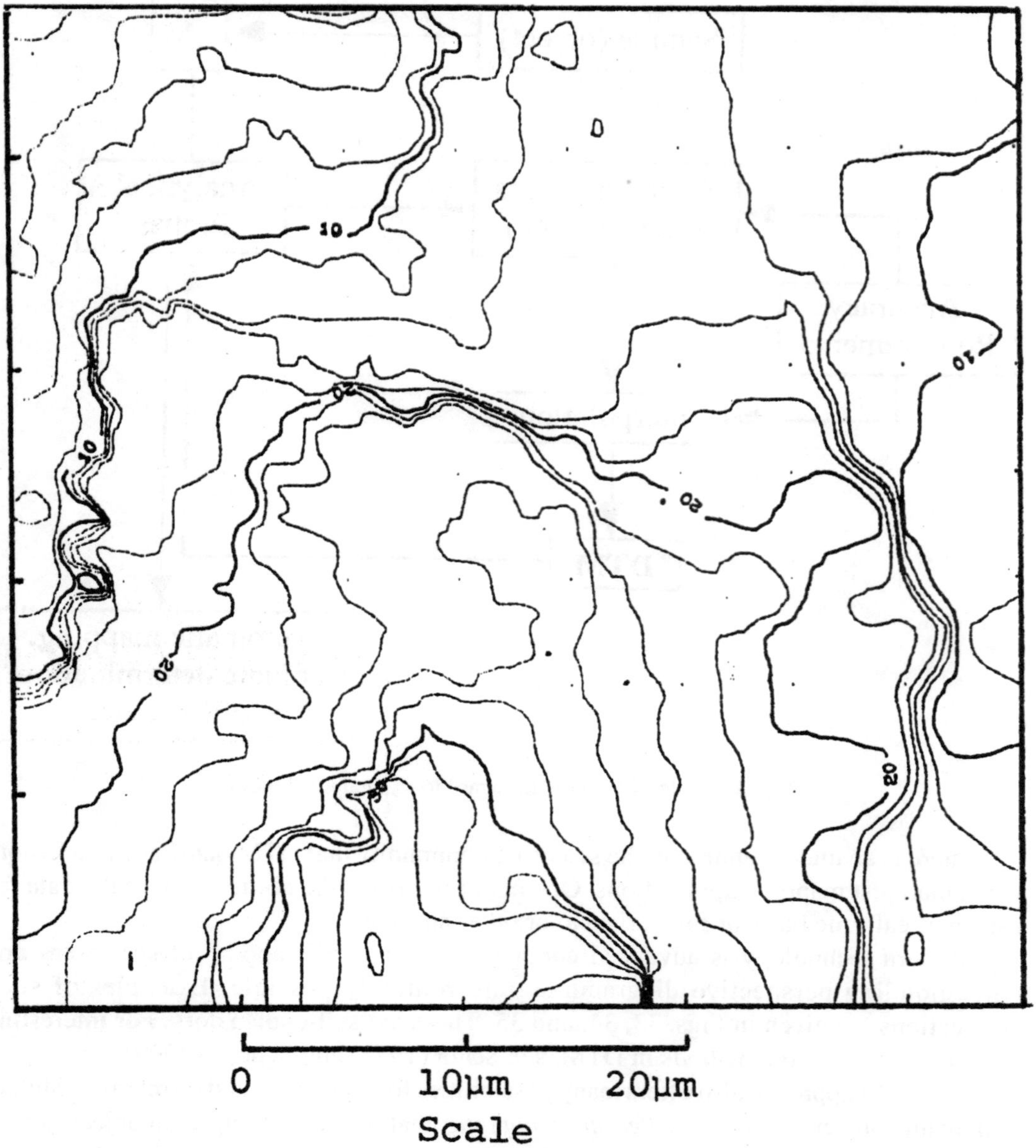

Fig. 33: Computer generated contour map with SEM micrographs
[Nickel-Copper High Strength steel surface: *Courtesy*: Adiguzel, 1985]

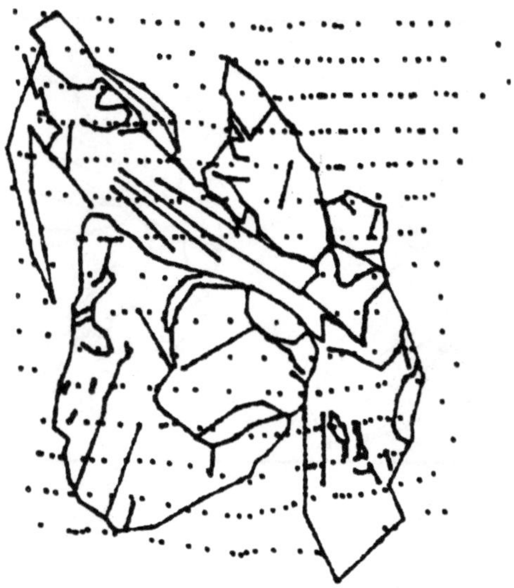

Fig. 34: Computer generated perspective view
[of the same sample as in Fig. 33]

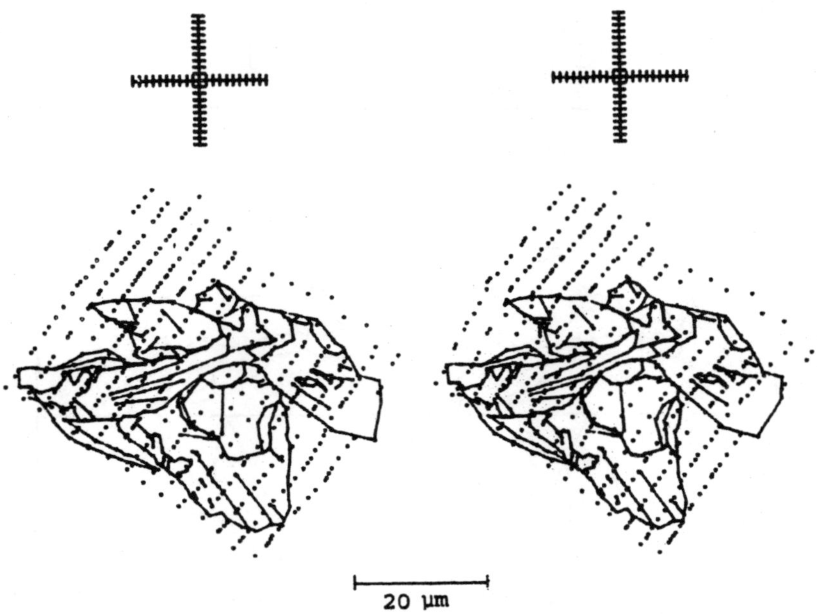

Fig. 35: Computer generated stereogram
[of the same sample as in Fig. 33]
Rotations: $\omega = \varphi = 0°; \kappa = 45°$

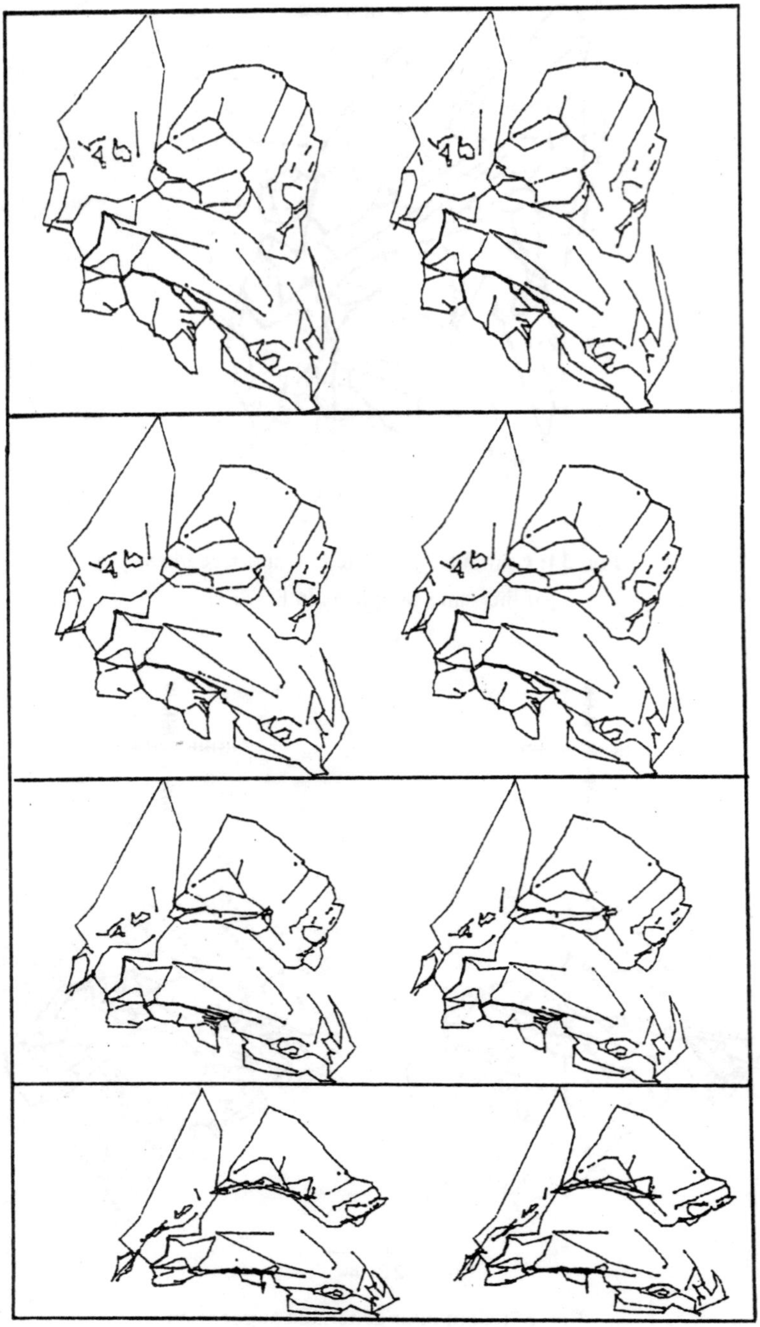

Fig. 36: Computer generated series of stereograms [of the same sample as in Fig. 33] by progressively changing Ω tilt (of model) by 15°

BIBLIOGRAPHY

Adiguzel, M. (1979), *Digital terrain models and their applications to perspective diagramming and contouring.* M.Sc. Thesis, The Ohio State University, Columbus, Ohio. vi + 149 pages.

Adiguzel, M. (1985). *Problems related to three-dimensional mapping with electron Micrographs.* Ph.D. Thesis, Lavel University, Quebec, Canada, xix + 229 pages.

American Society for Photogrammetry and Remote Sensing (2004). *Manual of Photogrammetry* (5th Ed.). ASPRS, Falls Church, Virginia. xvi + 1151 pages. Addendum (1996). *Digital Photogrammetry*: 250 pages. [Ed.: Maune, D.]

Ayeni, O. (1976). *Considerations for automated DTMs with applications in differential photo mapping.* Ph.D. Dissertation, The Ohio State University, Columbs, Ohio. viii + 199 pages.

Black, T.J. (1970). SEM, Scanning Electron Microscope. *Photographic Applications in Sciences, Technology and Medicine.*

Boyde, A. and Ross, H.F. (1975). Photogrammetry and the Scanning Microscope. *Photogrammetric Record,* 8 (46) : 408-457.

Eisenhart, C. (1969). *National Bureau of Standards (USA),* Spec. Publ. 300, Vol. 1.

El Chazali, M.S. (1978). *Some photogrammetric investigations of scanning and transmission electron micrography and their applications.* Ph.D. Dissertation, The Ohio State University, Columbus, Ohio. xiv + 215 pages.

Ghosh. S.K. (1971). Volume Determination with an Electron Microscope. *Photogrammetric Engineering,* 37 (2): 187-191.

Ghosh, S.K. (1972). *Theory of Stereophotogrammetry,* 2nd ed., Ohio State University Bookstores, Columbus, Ohio.

Ghosh, S.K. (1975). Photogrammetric Calibration of a Scanning Electron Microscope. *Photogrammetria,* 31 (3): 91-114.

Ghosh, S.K. and Nagaraja, H. (1976). Scanning Electron Micrography and Photogrammetry. *Photogrammetric Engineering and Remote Sensing,* 42 (5): 649-657.

Ghosh, S.K., El Ghazali, M.S. Deviney, M.L. and Mercer, H.N. (1978). *Three Dimensional Mapping by Combining Transmission and Scanning Electron Microscopes.* Int'l Society for Photogrammetry and Remote Sensing, Commission V Report (K. Torlegard, Ed.), pp 1-10.

Ghosh, S.K. and El Ghazali, M.S. (1977). *Stereo Electron Micrographic Studies of Carbon Black.* OSU Geodetic Science Report No. 261, The Ohio State University, Columbus, Ohio, vii + 52 pages.

Ghosh, S.K. and El Ghazali, M.S. (1979). *Application of Digital Terrain Models (DTM) to carbon black aggregates.* OSU Geodetic Science Report No. 283, The Ohio State University, Columbus, Ohio, iv + 24 pages.

Ghosh, S.K. and Adiguzel, M. (1984). Image rotation vis-a-vis magnification, a remarkable phenomenon in electron micrography. ISPRS XV Congress, Com. V, Rio-de-Janeiro, Brazil. *ISPRS Archives*, 25 (A5): 325-334.

Ghosh, S.K. (2005), *Foundamentals of Computational Photogrammetry*, Concept Publishing Company, New Delhi, India.

Helmcke, J.G. (1965). Determination of the third dimension of object by stereoscopy. *Quantitative Electron Microscopy*. Vol. 14 [Ed: Bahr, G.F. and Zeitler, E.H. US Armed Forces Institute of Pathology (Biophysics branch), Washington, D.C.

Hirvonen, R.H. (1971). *Adjustment by Least Squares in Geodesy and Photogrammetry*. Frederick Ungar Publ. Co., New York, N.Y.

Klemperer, O. and Barnett, M.E. (1971). *Electron Optics*, 3rd Ed. Cambridge university Press, Cambridge, U.K. xii + 506 pages.

Ladouceur, G., Trotier, P. and Allard, R. (1982). Zeiss Stereotop Modified into an Analytical Stereoplotter. *Photogrammetric Engineering and Remote Sensing*, 48(10): 1577-1580.

Leipholz, H. (1970). *Stability Theory*, Academic Press, New York.

Maune, D.F. (1973). *Photogrammetric Self-calibration of a Scanning Electron Microscope*. Ph.D. Dissertation, The Ohio State University, Columbus, Ohio. x + 227 pages.

Mikhail, E.M. (1976). *Observations and Least Squares*, IEP-A Dun-Donnelley Publisher, New York, N.Y.

Miller, C.L. and Laflamme, R.A. (1958). *Digital Terrain Model System Manual*, Mass. H.P.S. 1 (13). Massachusetts Department of Public Works and US Bureau of Public Roads, Boston, Mass.

Nagaraja, H. (1974). *Application studies of scanning electron microscope photographs for micro-measurements and 3-D mapping*. Ph.D. Dissertation, The Ohio State University, Columbus, Ohio. xviii + 181 pages.

Oatley, C.W. (1972). *The Scanning Electron Microscope, Part I – The Instrument*, The University Press, Cambridge, U.K.

Oshima, T., Kimoto, S. and Suganuma, T. (1970). Stereomicrography with a scanning electron microscope. *Photogrammetric Engineering and Remote Sensing*, 36(8): 874-879.

Schenk, T. (1999). *Digital Photogrammetry*, Terra Science, Laurelville, Ohio. xx + 428 pages.

Schut, G.H. (1977). *Review of interpolation methods for DTM*. ISPRS XXIII Congress, Helsinki, Finland. ISPRS Archives.

Thompson, M.M. and Rosenfield, G.H. (1971). On Map Accuracy Specifications. *Surveying and Mapping*, 31(1): 57-64.

Waldhausl, P. (1978). An approximate solution for the restitution of stereo electron micrographs. *Photogrammetric Engineering and Remote Sensing*, 44(8): 1005-1009.

Wells, O.C. (1960). *British Journal of Applied Physics*, 11: 199-201.

Wolf, P.R. and Ghilani, D. (1997). *Adjustment Computations - Statistics and Least Squares in Surveying and GIS*. Wiley Series in Surveying and Boundary Control. John Wiley & Sons, Inc., New York, N.Y. 564 pages.

Wood, R. (1972). The modification of a topographic plotter and its application in three-dimensional plotting of stereomicrographs, *Photogrammetric Record*, 7(40): 454-465.

INDEX